PHASE DIAGRAMS OF THE ELEMENTS

Phase Diagrams of the Elements

David A. Young

UNIVERSITY OF CALIFORNIA PRESS
Berkeley Los Angeles Oxford

University of California Press
Berkeley and Los Angeles, California

University of California Press
Oxford, England

Library of Congress Cataloging-in-Publication Data

Young, D. A. (David A.), 1942–
Phase diagrams of the elements/David A. Young
p. cm.
Includes bibliographical references and index.
ISBN 0-520-07483-1
1. Phase diagrams. 2. High pressure (Science) I. Title.
QD503.Y68 1991 90-25978
541.3'63—dc20 CIP

Printed in the United States of America
1 2 3 4 5 6 7 8 9

The paper used in this publication meets the minimum requirements of American National Standard for Information Sciences—Permanence of Paper for Printed Library Materials, ANSI Z39.48-1987 ♾

CONTENTS

PREFACE

In 1975 I wrote a short review article on the phase diagrams of the elements which was circulated as a Lawrence Livermore Laboratory Report. Since 1975, there has been an explosion of new experimental and theoretical research on high-pressure phase diagrams. Since nearly every element from hydrogen to einsteinium has now been studied at high pressure, this seems a good time to present a new review of the elemental phase diagrams. The main difference between this work and previous compilations is that I have now included a detailed discussion of theoretical calculations on phase changes and phase diagrams for the elements.

The material presented in this work is directed mainly toward condensed-matter physicists, but it may also be of interest to workers in the fields of earth and planetary sciences, astrophysics, metallurgy, and materials science. In view of a possible future edition of this book, I would welcome comments, corrections, and new data.

I have benefited enormously from discussions with my colleagues at Livermore and elsewhere. Special thanks are due to J. Akella, F. Bundy, R. Grover, W. Holzapfel, A. McMahan, J. Moriarty, H. Radousky, and D. Schiferl for comments on the manuscript. Also I thank K. Skinnell for the major task of preparing the figures for publication, and S. Kerst for preparation of the camera-ready manuscript.

David A. Young
Lawrence Livermore National Laboratory
Livermore, California 94550

CHAPTER 1
Introduction

This book is a response to the rapid development of experiment and theory in high-pressure condensed-matter physics. The diamond-anvil cell (DAC) has revolutionized the field. Experiments are now routinely performed above 100 GPa, and the first phase transitions above 100 GPa have recently been reported. There is also progress along the temperature axis, with high sample temperatures produced either by immersing a heat-resistant diamond-anvil assembly in a heat bath or by focusing laser light onto the sample and recording the temperature pyrometrically. This latter technique can be used up to the highest known melting points near 5000 K.

Theoretical methods have also been refined, and together with large supercomputers, they have been fashioned into increasingly accurate means of predicting solid structures and melting curves at high pressure. Theorists and experimentalists are now actively collaborating in the study of phase behavior at extremes of pressure and temperature, and the result has been a marked acceleration in the pace of research.

Most of the elements have now been studied in the DAC at high pressures, and numerous new crystal structures have been identified. The complex pattern of structures found in the elements is leading to a new understanding of chemical periodicity and of the old corresponding-states hypothesis. It is now clear that the structures of the solid and liquid elements reflect the full complexity of their atomic structures, and that a simple point-to-point mapping of one phase diagram onto another by a rigid application of the simple corresponding-states principle is incorrect.

The elements show a wide variety of unusual phenomena, such as insulator-metal transitions, electronic isostructural transitions, and melting-curve maxima and minima. It is therefore not necessary to consider compounds for the study of these phenomena, and the development of the theory is considerably simplified by focusing on the elements. There remains much theoretical work to be done.

Previous surveys of the phase diagrams of the elements have been compilations of experimental results[1–7], with minimal theoretical interpretation. In contrast, this book has been written with an emphasis on theory, and I have tried to show how the overall agreement between experimental data and theoretical models is now surprisingly good.

In chapters 2 and 3, I have summarized developments in experimental and theoretical methods, respectively. Chapters 4–15 are organized by the usual elemental groups of the periodic table, except that hydrogen is given a separate chapter. Each element is given a section where the experimental data and theoretical models are summarized. This is shown in Fig. 1.1. In order to cover the entire periodic table, I have summarized data for each element in telegraphic language, using abbreviations listed in Table 1.1. Where sufficient data are available for an element, one or more pressure-temperature phase diagrams are shown. The concluding section of each chapter is an attempt to develop general theoretical ideas about the group of elements in question.

In chapter 16 the low-pressure liquid-vapor transition is discussed for all of the elements. In chapter 17, there is a general discussion of the law of corresponding states and the periodic law in relation to the new high-pressure data for the elements. In the final appendixes, the important thermophysical properties of the elements are listed.

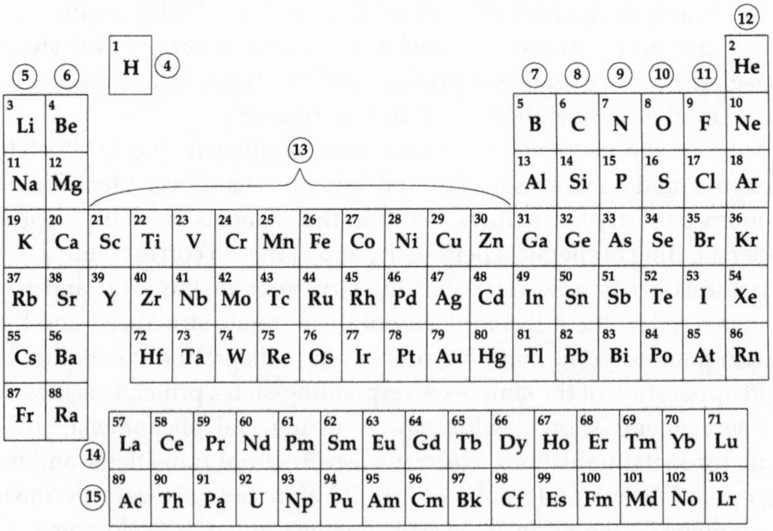

Fig. 1.1 Chapter organization of the book. Each circled chapter number is shown next to the group of elements described in that chapter.

TABLE 1.1 Abbreviations Used in This Book

AIP	*Ab Initio* Pseudopotential
APW	Augmented Plane Wave
ASA	Atomic-Sphere Approximation
ASW	Augmented Spherical Wave
ca.	circa
CDW	Charge-density wave
cp	close-packed
DAC	Diamond-Anvil Cell
DTA	Differential Thermal Analysis
EOS	Equation of State
FPLAPW	Full-Potential Linear Augmented Plane Wave
GPT	Generalized Pseudopotential Theory
KKR	Korringa-Kohn-Rostocker
LAPW	Linear Augmented Plane Wave
LCGTO	Linear Combination of Gaussian-Type Orbitals
LDA	Local-Density Approximation
LMTO	Linear Muffin-Tin Orbitals
LSDA	Local Spin-Density Approximation
LV	Liquid-Vapor
MC	Monte Carlo
MD	Molecular Dynamics
MT	Muffin Tin
NFE	Nearly Free Electron
NMR	Nuclear Magnetic Resonance
OCP	One-Component Plasma
QMC	Quantum Monte Carlo
RP	Room Pressure (1 atm)
RT	Room Temperature (300 K)
RTP	Room Temperature and Pressure (300 K and 1 atm)
XRD	X-Ray Diffraction
vs.	versus

Numerous references are provided to papers where the experiments and theories are described in detail. Nearly all of the high-pressure work cited in this book was done after 1960, and I have emphasized work done since 1980. I have cited papers either published or in preprint form up to September 1990. Statements like "the structure has not been determined" should be understood by the reader as a shorthand form of "as of September 1990, to the best of my knowledge the structure has not been determined." The references emphasize the latest research and are not intended to be complete. For older references, the reader should consult the reviews in Refs. 1–7.

First-order phase transitions are emphasized in this book. Magnetism, superconductivity, and charge-density waves are mentioned only when necessary, and no systematic discussion of these phenomena is offered.

Certain conventions are here adhered to for the construction of phase diagrams: 1) The pressure axis is vertical and the temperature axis is horizontal; 2) Temperature is in Kelvins and pressure is in gigapascals; 3) The origin is always $T = 0$, $P = 0$ so that the true size of the stability fields can be seen; 4) The phase boundaries are a synthesis, sometimes subjective, of different experimental results, but with emphasis on the most recent results; 5) Dashed lines indicate extrapolated or uncertain phase boundaries; and 6) Chaotic and inconsistent phase nomenclature has prompted me to use a shorthand crystallographic notation for solid structures which is shown in Table 1.2. The notation contains the Bravais lattice type together with the number of atoms (or molecules) in the conventional unit cell. The very common lattices bcc(2), fcc(4), and hex(2) are given the usual designations bcc, fcc, and hcp for clarity. The conventional unit cells are shown in Fig. 1.2. Phase diagrams are labeled with these notations as well as with the conventional symbol.

TABLE 1.2 Crystallographic Notation

Point group	Bravais lattice	Abbreviation	Example
cubic(cub)	simple cubic	sc	α-Po
	body-centered cubic	bcc	V
	face-centered cubic	fcc	Ar
tetragonal(tet)	simple tetragonal	st	γ-N_2
	centered tetragonal	ct	In
orthorhombic (orth)	simple orthorhombic	so	α-Np
	body-centered orthorhombic	bco	I_2 II
	face-centered orthorhombic	fco	γ-Pu
	base-centered orthorhombic	eco	Cl_2
monoclinic(mon)	simple monoclinic	sm	S I
	centered monoclinic	cm	Bi II
hexagonal	hexagonal	hex	Mg
rhombohedral	rhombohedral	rh	As
triclinic	triclinic	tr	—

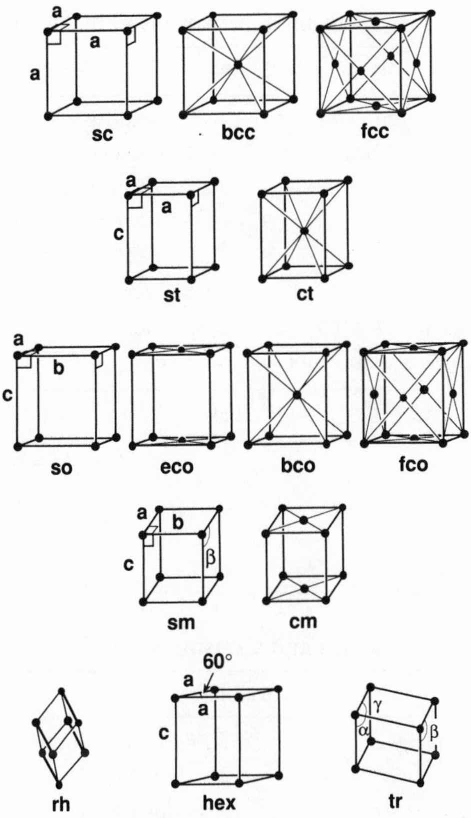

Fig. 1.2 The conventional unit cells for the 14 Bravais lattices. The abbreviation for each lattice type is shown.

I have used SI units where possible, but other units, such as the atomic units used in band-structure theories, are included in the text and graphics. A list of units used in the book is shown in Table 1.3. A list of the most commonly used variables and constants is shown in Table 1.4.

TABLE 1.3 Units and Prefixes Used in This Book

Units	
bohr; 1 bohr = 5.29×10^{-11} m	min = minute
eV = electron volt; 1 eV = 1.602×10^{-19} J; 1 eV/k = 11605 K	mol = mole
hr = hour	Pa = Pascal; 1 Pa = 1 Newton/m^2; 1 MPa = 10 bar; 1 GPa = 10 kbar; 1 TPa = 10 Mbar
J = Joule; 1 J = 10^7 erg	Ry = Rydberg; 1 Ry = 13.606 eV
K = Kelvin.	s = second
kg = kilogram; 1 Mg/m^3 = 1 g/cm^3	yr = year
m = meter; 1 m^3/Mmol = 1 cm^3/mol	
Prefixes	
n = nano = 10^{-9}	M = mega = 10^6
μ = micro = 10^{-6}	G = giga = 10^9
m = milli = 10^{-3}	T = tera = 10^{12}
k = kilo = 10^3	

TABLE 1.4 Variables and Constants Used in This Book

Variables	
A = Helmholtz free energy	R_A = ion- (or atomic-) sphere radius
E = energy	ρ = density
G = Gibbs free energy	S = entropy
H = enthalpy	T = temperature
P = pressure	Z = nuclear charge or atomic number
r_s = electron-sphere radius	Z_i = ionic charge or valence
Constants	
e = electron charge = 1.602×10^{-19} Coul	
h = Planck's constant = 6.626×10^{-34} J s	
$\hbar = h/2\pi$	
k = Boltzmann's constant = 1.381×10^{-23} J/K	
m_e = electron mass = 9.110×10^{-31} kg	
N = Avogadro's number = 6.022×10^{23} mol^{-1}	
R = gas constant = 8.314 J/mol K	

References

1. V. V. Evdokimova, Usp. Fiz. Nauk **88**, 93 (1966) [Sov. Phys. Usp. **9**, 54 (1966)].
2. W. Klement, Jr., and A. Jayaraman, Prog. Solid State Chem. **3**, 289 (1966).
3. J. F. Cannon, J. Phys. Chem. Ref. Data **3**, 781 (1974).
4. D. A. Young, Lawrence Livermore Laboratory Report UCRL-51902, September 1975.
5. C. W. F. T. Pistorius, Prog. Solid State Chem. **11**, 1 (1976).
6. E. Yu. Tonkov, *Fazoviye Diagrammy Elementov pri Vysokom Davleniya* [*Phase Diagrams of the Elements at High Pressure*] (Nauk, Moscow, 1979).
7. L.-G. Liu and W. A. Bassett, *Elements, Oxides, and Silicates: High-Pressure Phases with Implications for the Earth's Interior* (Oxford University Press, New York, 1986).

CHAPTER 2
Experimental Methods

2.1 Introduction

The experimental study of one-component phase diagrams requires the control and measurement of pressure and temperature as well as the ability to detect phase changes and to determine phase structures. Of these the principal technological challenge is the generation of ever-higher pressures.

There are three general experimental approaches to the study of phase diagrams. These are static high pressure, shock compression or dynamic high pressure, and isobaric heating at low ambient pressures. The paths of these methods in the *P*-*T* plane are shown in Fig. 2.1.

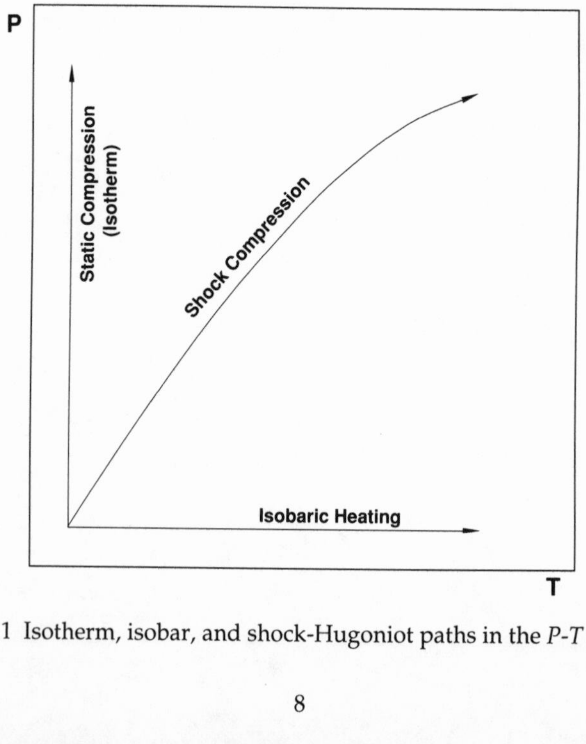

Fig. 2.1 Isotherm, isobar, and shock-Hugoniot paths in the *P*-*T* plane.

2.2 Static High Pressure

The effort to generate high static pressures has a long and interesting history[1,2]. Since the first experimental work in the 1700s, the rise of attainable pressure has been roughly exponential, reaching 0.1 GPa about 1830, 10 GPa around 1940, and 100 GPa in 1976. The basic problem of static high pressure is to design the high-pressure cell so that the sample is contained and plastic flow or fracture of the surrounding mechanical parts is avoided.

Below 2 GPa, the pressure can be applied to the sample cell by an external source of compressed gas[3]. The sample is contained in a cylindrical vessel connected by metal tubing to the pressure source. The pressure source might be a pump or an intensifier consisting of a cylinder of fluid with a piston for compression. The principal problems which must be solved in working with this apparatus are the prevention of leaks in the tubing junctions and the measurement of sample properties under pressure. Each measured property, such as volume, electrical resistance, melting point, or solid-solid phase transition, requires a different modification of the apparatus.

Control and measurement of temperature is a simple process of maintaining a constant-temperature heat bath around the sample chamber. The upper limit of temperature is determined by the mechanical strength of the sample-chamber material. Measurement of pressure is much more difficult. Primary pressure gauges are manometers based on the equation pressure = force/area. An example is a weighted piston balanced against a pressurized cylinder. This method is usable up to 2 GPa, but it is much more convenient to use secondary standards which depend on the variation of some physical property with pressure. Such standards must be calibrated against a believable primary standard at some point. A favorite secondary standard is the resistance of a manganin wire[4]. Manganin shows a linear dependence of resistance on pressure and is readily calibrated and reproducible. Nowadays 2 GPa is considered a low pressure for scientific research, but since virtually all industrial high-pressure processes operate in the 0–2 GPa range, this technology is very important[5].

Above 2 GPa, the fluid-pressure system fails because of the increasing viscosity or freezing of the fluid and the limited mechanical strength of the junctions joining the pressure tubing and the sample chamber. For higher pressures, therefore, the sample must be compressed directly by movable solid surfaces. The piston-cylinder apparatus does this by driving a piston into a hollow cylinder containing the sample[6,7].

The piston requires high compressive strength and is usually made of tungsten carbide. The cylinder requires high tensile strength and is made of high-strength steel. An unsupported, single cylinder design can be used up to about 3 GPa. With multiple cylinders for radial support, pressures up to 8 GPa are attainable. A sketch of a supported piston-cylinder apparatus is shown in Fig. 2.2.

The sample is encapsulated and surrounded with a mechanically weak medium, so that nearly hydrostatic conditions prevail. A small resistance heater can be introduced to maintain high sample temperatures. Primary pressure calibrations are done with the force/area relation, together with corrections for the deformation of piston and cylinder, and for friction. Secondary calibrations commonly use a series of room temperature (RT) phase transitions[8]. Phase transitions may be detected by electrical resistance measurements or by differential thermal analysis (DTA), which records the anomaly due to the latent heat of transition on the heat flow across the sample. Because of the massive parts lying between the

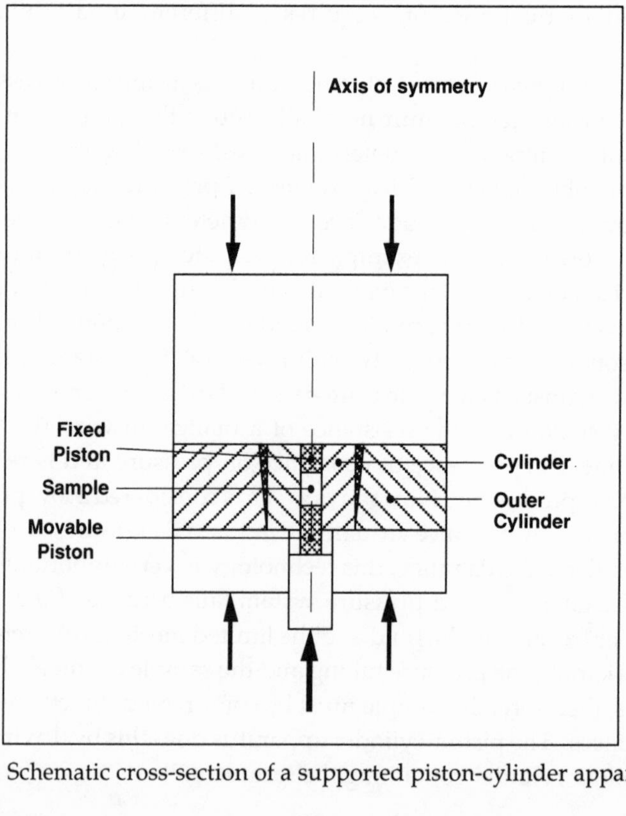

Fig. 2.2 Schematic cross-section of a supported piston-cylinder apparatus.

experimenter and the sample, the piston-cylinder apparatus is not convenient for optical or x-ray diffraction (XRD) measurements. The piston-cylinder apparatus is, however, an efficient means for mapping phase diagrams, and many of the diagrams described in this book were so determined.

For pressures above 8 GPa, the piston-cylinder apparatus suffers fracture of the piston, and other designs are needed. The opposed-anvil apparatus is now the most commonly used for pressures above 8 GPa[6]. Here two tapered pistons or anvils make contact at a very small surface area. This allows stresses larger than the compressive strength of the anvils to be achieved in the sample, which is compressed between the two anvils. The anvils are usually made of tungsten carbide or diamond.

Alternatives to the simple two-anvil configuration include the "belt" and "girdle" systems[6,9], which are hybrids of the piston-cylinder and opposed-anvil configurations, and multianvil configurations[9,10], in which four or more anvils are simultaneously brought together to compress a polyhedral sample volume. These high-pressure systems are able to generate pressures up to a few tens of GPa. They are more expensive and less widely used than the simple opposed-anvil design.

With diamond anvils, the opposed-anvil apparatus becomes the diamond-anvil cell (DAC), shown schematically in Fig. 2.3. The anvils are made by grinding the culet tips off of faceted gem-quality diamonds. The resulting anvil faces are 0.2 to 0.8 mm in diameter. To achieve a constant and nearly

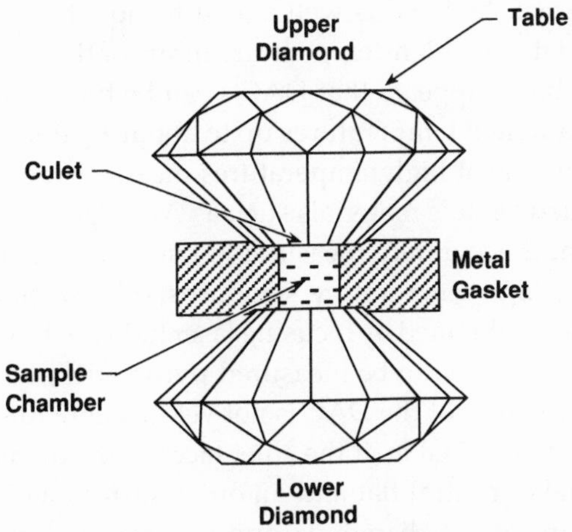

Fig. 2.3 Schematic diagram of diamond anvils with gasket and sample cell. (From Jayaraman[11]. Redrawn with permission.)

hydrostatic pressure on the sample, a gasket, which is a thin metal disk with a hole in the center, is placed between the anvils. The sample is a particle a few tens of micrometers in diameter, and is placed together with a mechanically weak medium in the gasket hole. The whole assembly is then squeezed between the anvils. With its low cost, very high attainable pressures, and easy optical or X-ray access to the sample, the DAC is the most powerful and versatile high-pressure apparatus ever invented[11,12]. DAC technology has spread rapidly and now dominates static high-pressure research.

Primary pressure calibration in anvil devices is very difficult because of the very small contact area and the typically nonhydrostatic and nonuniform state of stress across the sample. The technique now common is to use the equation of state of various cubic metals as the primary pressure standard[13]. Pressure-volume RT isotherms of metals such as Cu, Au, Pt, and W are derived from absolute experimental shock-wave pressures by means of the Grüneisen model[13]. The volumes of the metals measured by XRD in the DAC may then be correlated with the pressure in the gasket hole. A very useful secondary standard is the shift in the R_1 fluorescence line of ruby. The pressure is found to be a smooth function of the wavelength change $\Delta\lambda$[13]. At present, the ruby standard has been extended to 180 GPa[14]. Above 100 GPa, however, fluorescence from the diamond anvils increasingly obscures the ruby line, and the use of simple metals together with XRD as the pressure standard is the most reliable way to obtain the pressure. At the time of this writing, the highest pressure obtained by this method was 364 GPa[15].

The small size of the DAC is well suited for low-temperature experiments in a cryostat. The high-temperature limits of the apparatus are still being explored, but it appears that DAC's can be built with special alloy parts to sustain ambient temperatures up to about 1500 K[16]. The broadening of the ruby line at high temperatures means that another pressure standard is needed. Cubic materials such as Au, MgO, and W, with high melting points and accurately measurable equations of state have been recommended as high-temperature standards[16]. Much higher sample temperatures can be obtained by focusing laser light on the sample, and the steady-state temperature can be measured pyrometrically[17]. Likewise, the upper pressure limit of the DAC is not yet in sight. Routine measurements are made up to 50 GPa. If the culet faces of the diamond anvils are cut to give a smaller central flat and an outer surface with a small bevel angle, the fracture of the diamonds due to "cupping" is avoided, and pressures above 100 GPa can be achieved. Claims of pressures up to

550 GPa[18,19] are controversial, however, because of the problem with the ruby scale at very high pressures. It is now possible to achieve the highest pressures by optimizing the shape of the diamonds and the gasket characteristics with continuum-mechanics simulation codes[19,20].

Diagnostics for detecting phase changes in the DAC are numerous. The most general is XRD. In the DAC, the X-ray beam can be passed either through the diamonds or through a beryllium gasket. Two configurations of the DAC for powder diffraction are shown in Fig. 2.4[21]. The angle-dispersive technique (Fig. 2.4a) uses monochromatic X-rays and a photographic film for recording. The diffracted radiation appears as lines of varying intensity on the exposed film. The Bragg angle can then be calculated from $\tan(2\theta_{hkl}) = x_{hkl}/s$. Because of the small sample size and because the photographic film is not an efficient photon collector, very long exposure times (days) are required. The process can be speeded up with more efficient detectors. The energy-dispersive technique (Fig. 2.4b) uses incident white radiation and a solid-state detector at a fixed angle 2θ. The detector can discriminate photons of different energy and thus record a spectrum. The peaks at energies E_{hkl} are related to the Bragg angle by $hc/E_{hkl} = 2d_{hkl}\sin\theta$. A disadvantage of this method is the appearance of unwanted fluorescence peaks in the spectrum.

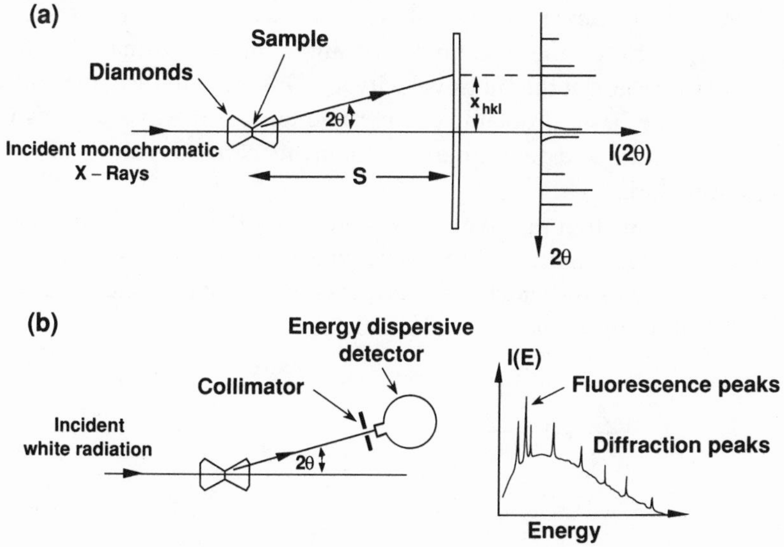

Fig. 2.4 Typical DAC configurations for powder X-ray diffraction: (a) angle-dispersive; (b) energy-dispersive. (From Menoni and Spain[21]. Redrawn with permission.)

The principal experimental constraint on XRD in the DAC is the long exposure time required. New techniques, such as the use of high-intensity synchrotron radiation, have reduced this time dramatically. XRD can now be used to measure the pressure, to detect phase changes, and to determine the structures of new phases.

The physical properties of diamond allow a multitude of spectroscopic measurements on the DAC sample chamber which can be useful in phase diagram studies. Examples are Raman, Brillouin, Mössbauer, and NMR spectroscopy and optical absorption and reflection measurements[11,12]. Other measurements of electrical resistance[11,12], positron annihilation[12], sound speed[22], and neutron scattering[23] are also of interest in the study of phase transitions.

2.3 Dynamic High Pressure

The study of matter compressed by shock waves, having begun during World War II, has a shorter history than static compression, but it has reached a comparable level of experimental precision and theoretical understanding[24–26]. A shock wave is a disturbance which propagates at supersonic speed in the medium. We can imagine the shock as arising from a piston which moves at a constant speed u_p into the medium. The boundary between the uncompressed and compressed material moves ahead of the piston at the shock velocity u_s. This is shown schematically in Fig. 2.5. The basic problem of experimental shock-wave physics is to measure u_p and u_s and determine from them the thermodynamic state of the shocked material.

The Rankine-Hugoniot equations determine the state of the shocked material through equations expressing conservation of mass, momentum, and energy across the shock front[24]. Assuming that the material has no mechanical strength, these equations are

$$\frac{V}{V_o} = 1 - \frac{u_p}{u_s} \tag{2.1}$$

$$P - P_o = \frac{u_s u_p}{V_o} \tag{2.2}$$

$$E - E_0 = \frac{1}{2}(P + P_0)(V_o - V) \tag{2.3}$$

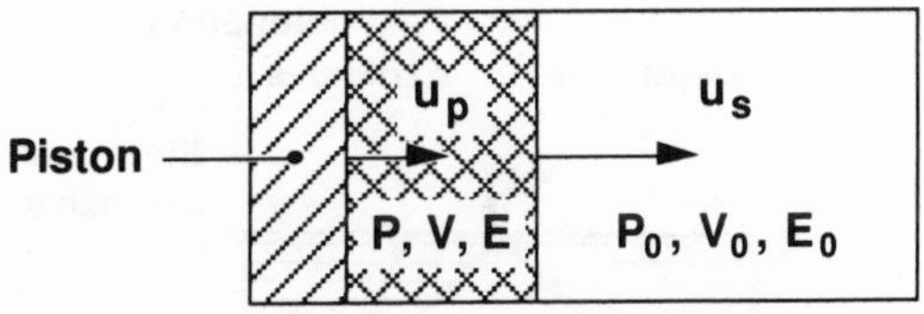

Fig. 2.5 A one-dimensional shock wave is generated in a medium (initially at P_0, V_0, E_0) by a piston moving at velocity u_p. The shock wave moves at velocity u_s and shocks the material to state (P, V, E).

where V = volume, P = pressure, E = energy, and the subscript "0" refers to the initial state. The locus of (P,V) points that can be reached by shock compression from a given initial state is called the Hugoniot.

There are two basic experimental techniques for generating shock waves, high explosives[24] and gas guns[27]. In the first case, illustrated in Fig. 2.6, a high-explosive lens system is shaped to produce a planar detonation wave. This wave launches an impactor into the sample. The two important variables to measure are the impactor velocity u_d at the time of collision, and the shock velocity u_s in the sample. The shock velocity can be measured either with grooves in the sample of varying depth which allow optical detection of the arrival time of the shock, or with self-shorting pins at varying depths which give an electrical signal at shock-arrival time.

The two-stage light-gas gun is sketched in Fig. 2.7. Here a propellant drives a heavy piston into hydrogen gas, compressing it strongly. At maximum compression a rupture valve bursts and the compressed hydrogen propels an impactor down the barrel and into the target chamber. With flash X-ray radiography the velocity of the impactor can be measured accurately, and with large arrays of shorting pins, the shock velocity in the target can be determined accurately. Although the gas gun is a more powerful and precise instrument than the explosive system, it remains true that most of the published Hugoniot data have been generated with high explosives[28,29].

If the impactor and target are made of the same material, then by symmetry $u_p = u_d/2$. With this symmetric-impact technique, the Hugoniots of standard materials such as Al, Cu, and Ta can be determined. The impedance-matching technique[27], which makes use of the equality of the u_p's and P's on both sides of the impact surface, is then used to determine the Hugoniots of nonstandard materials.

It is a useful empirical fact that for most materials the relation between u_s and u_p is linear. When the Hugoniot crosses a phase line, however, this relationship may no longer hold, and the u_s-u_p plot may show a discontinuity[24]. This happens because a steady shock wave requires a

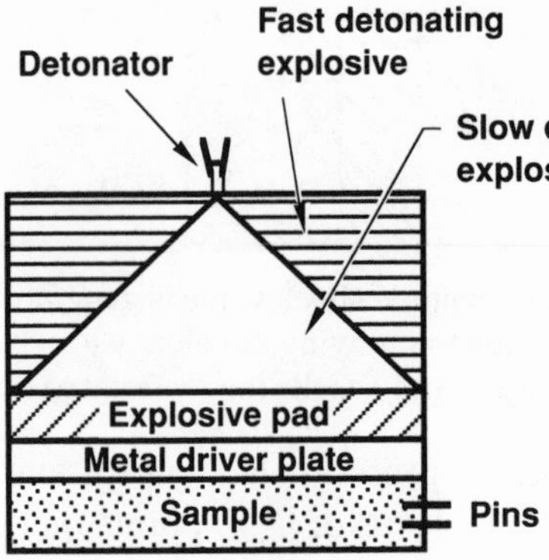

Fig. 2.6 High-explosive shock generator.

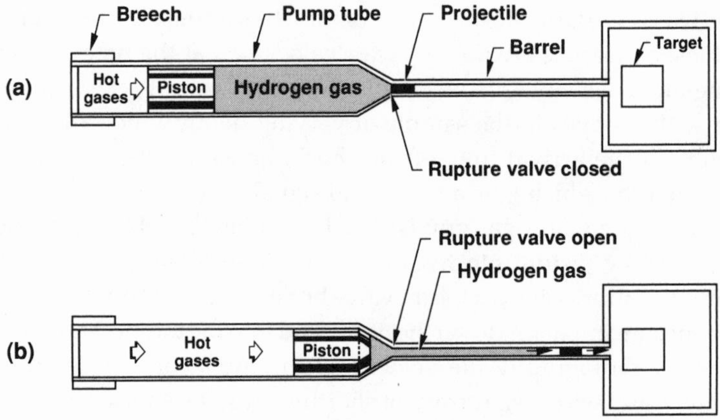

Fig. 2.7 Two-stage light-gas gun: (a) stage 1, gas compression; (b) stage 2, impactor acceleration.

sound speed that increases with compression. A first-order phase transition violates this requirement, with the result that the wave breaks up into two parts. The low-pressure wave represents the untransformed material, and the high-pressure wave the transformed material. Since the standard detectors will only register the arrival of the first (low-pressure) wave, this two-wave region appears as a flat segment of constant u_s on the u_s-u_p plot. When the shock velocity in the transformed material rises above that of the untransformed material, a third segment appears on the u_s-u_p plot. The behavior of the shock Hugoniot as it crosses a phase boundary is illustrated in Fig. 2.8.

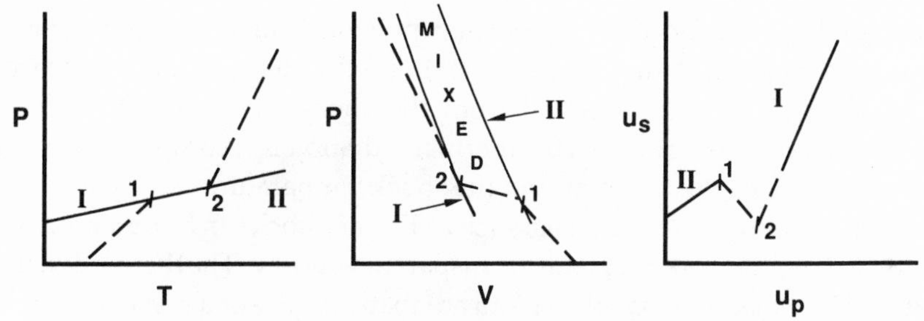

Fig. 2.8 Path of shock Hugoniot across a phase boundary. (From McQueen, et al.[24]. Redrawn with permission.)

The appearance of discontinuities in the experimental u_s-u_p plane is a good indication of a phase transition. However, if the P-T slope of the phase line is not too different from that of the Hugoniot, as is usually true with melting, the only effect on the u_s-u_p plot will be a nearly undetectable change of slope. Given the limited precision of shock-wave measurements, this is not a useful procedure for locating the melting curve.

A new "optical-analyzer" technique of high sensitivity has been developed for detecting phase transitions along the shock Hugoniot to arbitrarily high pressures[30]. Here a target is machined with a series of steps on the back side and this surface is placed in contact with a dense liquid such as bromoform. A very thin impactor launched either by explosives or gas gun strikes the front surface of the target and shocks move forward into the target and backward into the impactor. Because the impactor is very thin, the backward-moving shock quickly reaches the free surface and is replaced by a faster, forward-moving rarefaction wave moving at the longitudinal sound speed of the shocked material. The target is designed so that this wave will finally overtake the forward-moving shock at some distance into the liquid. This event can be observed as a sharp diminution of the measured optical-radiation intensity coming from the advancing shock front. The different thicknesses of target material in the steps lead to different overtaking times, and these data can be extrapolated to obtain the thickness of target material corresponding to the overtaking occurring exactly at the surface. The sound speed in the shocked target material is then easily computed. Since the sound speed is a sensitive function of the crystal structure, phase transitions will be seen as discontinuities in the plot of sound speed vs. Hugoniot pressure.

Other shock-wave measurements with relevance to phase diagrams are flash X-ray diffraction[31] and pyrometric temperature measurements[32]. In flash X-ray diffraction, the crystal structure of the shocked single-crystal material is determined by a very short pulse of X-rays. In the pyrometric measurements the temperature of the shocked material is determined from the thermal radiation emitted from the shock front. For transparent materials such as liquid N_2, the thermal radiation is recorded as the shock moves through the material, and the shock temperature can be calculated directly. For opaque materials, such as Fe, the shock can be observed only as it arrives at an interface with a transparent window. The thermodynamic state of the material may then be an adiabatic release or a reshock from the original shock state, depending on the impedance mismatch at the interface. The analysis of the temperature is much more complex in this case[33]. When the shock Hugoniot crosses a first-order transition, as shown in Fig. 2.8, there is a marked change in the slope of the *P-T* path. Pyrometric measurements can therefore be used to locate phase transitions.

2.4 Isobaric Heating

In isobaric heating the thermodynamic properties and the crystal structures of a material are determined as a function of temperature along an isobar. The isobar pressure is typically low, from 0.1 to 500 MPa. The most commonly measured variables are volume, enthalpy, and sound speed as functions of temperature. These variables will show discontinuities at first-order phase transitions and are thus useful indicators of the existence of transitions. If the values of ΔV and ΔH across the transition can be determined accurately, then the slope of the phase boundary can be calculated from the Clausius-Clapeyron equation, $dP/dT = \Delta H/T\Delta V$. XRD may be used to determine the structures of any new phases, as well as their volumes. Static measurements up to the melting point along the 0.1 MPa isobar have been carried out for all but the most radioactive of the elements, and for some, these are still the only data available on the phase diagram[34,35].

For refractory materials, the determination of the ΔH and ΔV of melting cannot be done statically because no container with sufficiently high melting point is available. For such materials which are electrically conducting, a dynamic resistive-heating method is available[36,37]. Here the material to be studied is made into a wire and is enclosed in an inert-gas environment. A capacitor bank is discharged through the sample with a pulse roughly 50 μs long. This is long enough so that thermodynamic equilibrium prevails in the sample but short enough so that the column of liquid will not be significantly perturbed by gravity. The enthalpy can be determined by integrating the product of the measured current and voltage over time, and the volume can be determined by a streak-camera record. The temperature is measured pyrometrically. This apparatus is sketched in Fig. 2.9.

For measurements out to the liquid-vapor boundary and the critical region, the sample must be contained at pressures up to a few hundred MPa[36]. This is done with a pressure vessel fitted with optical windows for the volume and temperature measurements, as indicated in Fig. 2.9. Many useful data on melting and liquid thermodynamics have been collected by this method. So far, the highest temperature achieved in this apparatus is about 8000 K.

2.5 Discussion

We are presently in the midst of a period of rapid growth in experimental high-pressure physics. The possibilities of the DAC are still being explored, and we can expect not only further increases in the routinely achievable pressures but very great increases in the achievable sample temperatures. The DAC has become, and is likely to continue to be, the most important experimental tool for exploring high-pressure phase diagrams. The upper pressure limit of the DAC will probably be set by the first phase transition from diamond to a denser phase, because this is likely to lead to mechanical failure of the anvils. Theoretical calculations indicate that this may occur near 1.1 TPa[38].

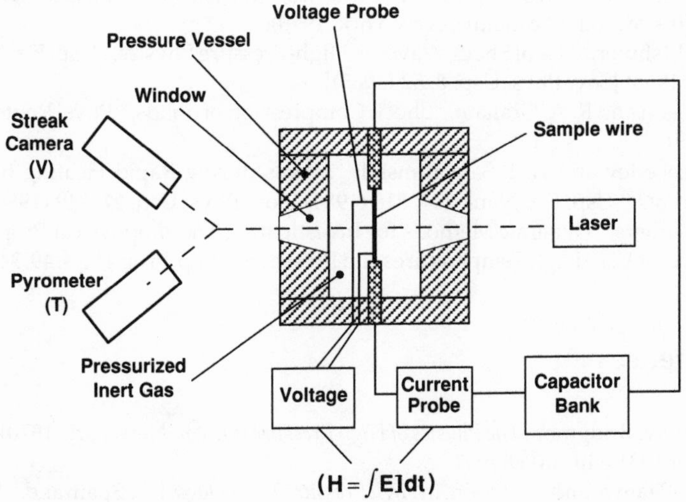

Fig. 2.9 Schematic diagram of the isobaric-heating apparatus.

Shock-wave techniques have also steadily progressed in precision and in the number of measurable variables. The optical-analyzer sound-speed measurement technique is the only method presently available for detecting phase transitions at pressures above 100 GPa and very high temperatures. At shock pressures above 500 GPa, even the most refractory metals have melted, and further compression will take place in the liquid. Thus the principal interest in the shock-wave method is for the location of phase transitions below this limit.

At low pressures and high temperatures, the isobaric-heating method promises to provide more data on the refractory elements. The determination of very-high-temperature critical points is likely to be done by the dynamic isobaric-heating method.

For Further Reading

P. W. Bridgman, *The Physics of High Pressure* (Dover, New York, 1970) (reprint of 1931 edition).

C. C. Bradley, *High Pressure Methods in Solid State Research* (Butterworths, London, 1969).

G. N. Peggs, ed., *High Pressure Measurement Techniques* (Applied Science Publishers, London, 1983).

A. Jayaraman, "Diamond Anvil Cell and High-Pressure Physical Investigations," Revs. Mod. Phys. **55**, 65 (1983).

A. Jayaraman, "Ultrahigh Pressures," Rev. Sci. Instrum. **57**, 1013 (1986).

R. G. McQueen, S. P. Marsh, J. W. Taylor, J. N. Fritz, and W. J. Carter, "The Equation of State of Solids from Shock Wave Studies," in *High-Velocity Impact Phenomena*, R. Kinslow, ed. (Academic, New York, 1970).

L. V. Al'tshuler, "Use of Shock Waves in High-Pressure Physics," Usp. Fiz. Nauk **85**, 197 (1965) [Sov. Phys. Usp. **8**, 52 (1965)].

L. Davison and R. A. Graham, "Shock Compression of Solids," Phys. Repts. **55**, 255 (1979).

S. V. Lebedev and A. I. Savvatimskii, "Metals during Rapid Heating by Dense Currents," Usp. Fiz. Nauk **144**, 215 (1984) [Sov. Phys. Usp. **27**, 749 (1984)].

G. R. Gathers, "Dynamic Methods for Investigating Thermophysical Properties of Matter at Very High Temperatures and Pressures," Rep. Prog. Phys. **49**, 341 (1986).

References

1. P. W. Bridgman, *The Physics of High Pressure* (Dover, New York, 1970) (reprint of 1931 edition) chap. 1.
2. J. Paauwe and I. L. Spain, in *High Pressure Technology*, I. L. Spain and J. Paauwe, eds. (Marcel Dekker, New York, 1977) vol. 1, chap. 1.

3. Ref. 1, chap. 2.
4. Ref. 1, chap. 3.
5. Ref. 2, vol. 2.
6. B. Crossland and I. L. Spain, in *High Pressure Measurement Techniques*, G. N. Peggs, ed. (Applied Science Publishers, London, 1983) chap. 8.
7. I. L. Spain, in Ref. 2, vol. 1, chap. 11.
8. V. E. Bean, in Ref. 6, chap. 3.
9. C. C. Bradley, *High Pressure Methods in Solid State Research* (Butterworths, London, 1969) chap. 6.
10. A. Onodera, High Temp.-High Press. **19**, 579 (1987).
11. A. Jayaraman, Revs. Mod. Phys. **55**, 65 (1983).
12. A. Jayaraman, Rev. Sci. Instrum. **57**, 1013 (1986).
13. H. K. Mao, P. M. Bell, J. W. Shaner, and D. J. Steinberg, J. Appl. Phys. **49**, 3276 (1978).
14. P. M. Bell, J. Xu, and H. K. Mao, in *Shock Waves in Condensed Matter*, Y. M. Gupta, ed. (Plenum, New York, 1986) p. 125.
15. Y. K. Vohra, H. Xia, H. Luo, and A. L. Ruoff, Appl. Phys. Lett. **57**, 1007 (1990).
16. D. Schiferl, J. N. Fritz, A. I. Katz, M. Schaefer, E. F. Skelton, S. B. Qadri, L. C. Ming, and M. H. Manghnani, in *High Pressure Research in Mineral Physics*, M. H. Manghnani, Y. Syono, eds. (KTK Scientific, Tokyo, 1987) p. 75.
17. D. L. Heinz and R. Jeanloz, in Ref. 16, p. 113.
18. J. A. Xu, H. K. Mao, and P. M. Bell, Science **232**, 1404 (1986).
19. W. C. Moss, J. O. Hallquist, R. Reichlin, K. A. Goettel, and S. Martin, Appl. Phys. Lett. **48**, 1258 (1986).
20. W. C. Moss and K. A. Goettel, Appl. Phys. Lett. **50**, 25 (1987).
21. C. S. Menoni and I. L. Spain, in Ref. 6, chap. 4.
22. J. M. Brown, L. J. Slutsky, K. A. Nelson, and L.-T. Cheng, Science **241**, 65 (1988).
23. V. P. Glazkov, S. P. Besedin, I. N. Goncharenko, A. V. Irodova, I. N. Makarenko, V. A. Somenkov, S. M. Stishov, and S. Sh. Shil´shtein, Pis´ma Zh. Eksp. Teor. Fiz. **47**, 661 (1988) [Sov. Phys. JETP Lett. **47**, 763 (1988)].
24. R. G. McQueen, S. P. Marsh, J. W. Taylor, J. N. Fritz, and W. J. Carter, in *High-Velocity Impact Phenomena*, R. Kinslow, ed. (Academic, New York, 1970) chap. 7.
25. L. V. Al´tshuler, Usp. Fiz. Nauk **85**, 197 (1965) [Sov. Phys. Usp. **8**, 52 (1965)].
26. L. Davison and R. A. Graham, Phys. Repts. **55**, 255 (1979).
27. W. J. Nellis, in Ref. 6, chap. 2.
28. M. van Thiel, ed., *Compendium of Shock Wave Data*, Lawrence Livermore Laboratory Report UCRL-50108, 1977.
29. S. P. Marsh, ed., *LASL Shock Hugoniot Data* (University of California Press, Berkeley, Los Angeles, London, 1980).
30. R. G. McQueen, J. W. Hopson, and J. N. Fritz, Rev. Sci. Instrum. **53**, 245 (1982).
31. Q. Johnson, A. C. Mitchell, and I. D. Smith, Rev. Sci. Instrum. **51**, 741 (1980).
32. G. A Lyzenga, T. J. Ahrens, W. J. Nellis, and A. C. Mitchell, J. Chem. Phys. **76**, 6282 (1982).
33. H. Tan and T. J. Ahrens, High Press. Res. **2**, 159 (1990).

34. Y. S. Touloukian and E. H. Buyco, eds., *Thermophysical Properties of Matter* (Plenum, New York, 1970), vol. 4 (Specific Heat) and vol. 12 (Thermal Expansion).
35. R. Hultgren, P. D. Desai, D. T. Hawkins, M. Gleiser, K. K. Kelley, and D. D. Wagman, *Selected Values of the Thermodynamic Properties of the Elements* (Am. Soc. Metals, Metals Park, Ohio, 1973).
36. G. R. Gathers, Rep. Prog. Phys. **49**, 341 (1986).
37. S. V. Lebedev and A. I. Savvatimskii, Usp. Fiz. Nauk **144**, 215 (1984) [Sov. Phys. Usp. **27**, 749 (1984)].
38. S. Fahy and S. G. Louie, Phys. Rev. B **36**, 3373 (1987).

CHAPTER 3
Theoretical Methods

3.1 Introduction

In order to make theoretical predictions of phase transitions we must first construct good theories of the free energies of the solid and liquid phases. It is not yet possible to do this completely from first principles, and instead a chain of approximations is required to obtain the free-energy functions needed for the construction of a phase diagram. Nevertheless, the theory has achieved a surprisingly high level of accuracy, and many phase transitions have been computed and found to be in good agreement with experiment.

The starting point for the study of condensed matter is electron-band-structure theory, which provides the ground-state energy of an assembly of nuclei and electrons. This theory is in principle capable of starting only with the atomic number Z and then predicting metallic, ionic, covalent, and insulating van der Waals types of solid-state bonding. In order to extend the calculations to nonzero temperature, we must generate accurate interatomic potentials. These, together with accurate statistical-mechanical theories of solids and liquids, then yield free energies and predictions for the phase diagram.

3.2 Electron-Band-Structure Calculations

One of the central problems of solid-state physics is to explain how free atoms can condense into a stable crystalline state. This is a formidable problem in quantum many-body theory, and it is only with the advent of supercomputers that accurate computations have been possible. In a truly *ab initio* calculation, one begins with a periodic array of nuclei of charge Ze together with Z electrons per nucleus, and then solves the Schrödinger equation for the total energy of the system.

It is now possible to attack this problem for low-Z atomic solids by means of Quantum Monte Carlo (QMC) methods[1], which are exact in principle. In QMC the Schrödinger equation is written as a diffusion equation in imaginary time. The N-body system is represented by an ensemble of boxes, each with N particles. The initial distribution of the particles is determined by an approximate trial wave function. Each system is considered in turn, and the particles are randomly displaced according to the "diffusion constant" $\hbar^2/2m_e$. Then "birth" or "death" of the system is determined by the change in potential energy resulting from the random move. The trial energy is slowly varied to stabilize the ensemble population, and the energy value required to achieve stability is the true ground-state energy. Similarly, the particle distribution in the ground state is a sampling of the true ground-state wave function.

Problems arise when the potential energy becomes large and negative as with electrons and nuclei, and when fermions are simulated. The first problem has been solved by importance sampling, which advects particles away from low-probability configurations. The second problem arises from the regions of negative-valued wave functions required by fermion antisymmetry. The problem of locating the correct nodes of the wave function is still not satisfactorily solved. So far, the QMC method has been applied successfully to condensed H_2, He, and Li, but accurate treatment of heavier atoms will require new techniques, such as the use of pseudopotentials.

For routine solid-state calculations, QMC is not yet practical, and strategic approximations must be made. The objective of modern electron-band-structure theory is to compute the total energy of a system of electrons moving in the field of a fixed lattice of nuclei[2]. This theory makes use of the one-electron approximation and the local-density approximation (LDA). This means that the full many-body wave function is approximated as a product of one-electron functions, and that the exchange-correlation energy is written as a functional of the local electron density, $n(\boldsymbol{r})$, where

$$n(\boldsymbol{r}) = \sum_k |\Psi_k(\boldsymbol{r})|^2 , \tag{3.1}$$

and $\Psi_k(\boldsymbol{r})$ is the one-electron wave function for occupied state k.

According to density-functional theory[3], the total energy of a system of nuclei and electrons is a unique functional of $n(\boldsymbol{r})$, and is a minimum at the true ground state. The total energy may be written:

$$E_{tot} = E_K + E_{en} + E_{ee} + E_{xc} + E_{nn} , \tag{3.2}$$

where the terms on the right are the kinetic, electron-nucleus, electron-electron, exchange-correlation, and internuclear energies. All terms but the last are functionals of $n(\mathbf{r})$. The unknown exact exchange-correlation functionals may depend on values of $n(\mathbf{r}')$ at locations $\mathbf{r}' \neq \mathbf{r}$, but the LDA assumes that this term is a function of $n(\mathbf{r})$ at $\mathbf{r}$ only[3].

The one-electron Schrödinger equation is then written

$$\left[-\frac{\hbar^2}{2m_e}\nabla^2 + W(\mathbf{r}) + \frac{e^2}{2}\int d\mathbf{r}' \frac{n(\mathbf{r}')}{|\mathbf{r}-\mathbf{r}'|} + V_{xc}[n(\mathbf{r})]\right]\Psi_k(\mathbf{r}) = \varepsilon_k \Psi_k(\mathbf{r}) , \tag{3.3}$$

where $W(\mathbf{r}) = \sum_i w(|\mathbf{r} - \mathbf{R}_i|)$, $\mathbf{R}_i$ represents the nuclear positions, and w is the nuclear Coulomb potential in an all-electron calculation or a pseudopotential otherwise. The general form of the exchange-correlation potential V_{xc} is not known, but a number of approximations based on the homogeneous electron gas, such as those of Slater[4] and Hedin and Lundquist[5], are in use. Eqs. (3.1) and (3.3) are solved self-consistently.

In solving Eq. (3.3), the periodicity of the lattice (Bloch condition) must be imposed as a boundary condition. The different methods used to solve the Schrödinger equation and to satisfy this condition have given rise to a variety of band-structure methods, some of which are listed in Table 3.1.

The Linear Muffin-Tin Orbital (LMTO) theory[6] is based on a number of further approximations. The muffin-tin (MT) potential assumes that within

TABLE 3.1 Local-Density Band-Structure Methods

Name	Acronym	Comments	Shape approx?
Ab initio pseudopotential	AIP	pseudopotential	no
Linear muffin-tin orbital	LMTO	all electron	yes
Augmented spherical wave	ASW	all electron	yes
Linearized augmented plane wave	LAPW	all electron	yes
Full-potential LAPW	FPLAPW	all electron	no
Linear combination of Gaussian-type orbital	LCGTO	all electron	no

a sphere inscribed in the primitive cell the atomic potential $V(\mathbf{r})$ is spherically symmetric, and that in the interstitial region $V(\mathbf{r})$ is constant. In LMTO, the MT is further simplified with the atomic-sphere approximation (ASA), in which the spherical potential is extended to the full atomic volume, so that the net interstitial volume is zero.

The Bloch condition is imposed by requiring cancellation of all neighbor wave functions within the atomic sphere. The "L" in LMTO refers to the approximation that the basis functions are made energy-independent, which allows the eigenfunctions to be obtained in a single diagonalization step. This speeds up the calculation by a factor of about 100, and is crucial to the success of the method. LMTO is suitable for heavy elements with large electron angular momentum (l), and the ASA permits the total pressure to be broken down into l-dependent pieces, which is useful for analysis of trends in the periodic table. The principal limitation of LMTO is the restriction to high-symmetry crystal structures imposed by the ASA.

In the *ab initio* Pseudopotential (AIP) theory[7], the valence electron-ion interaction is represented by l-dependent (nonlocal) pseudopotentials which have been fitted to free-atom energy levels. The Schrödinger equation, Eq. (3.3), is recast in momentum variables, and the wave functions are expanded in plane waves which automatically satisfy the Bloch condition. No muffin-tin or other shape approximation is used.

For *sp*-electron-bonded solids, this method is very accurate and can represent covalently-bonded open structures as well as metallic solids. For more localized *d*- and *f*-orbitals, a mixed basis set and more intensive computing are required. Also, the pseudopotential fails at pressures where the ion cores overlap.

Effects due to the relativistic motion of the electrons are very small in light atoms, but they increase roughly as Z^2, and they must be included in calculations on heavy elements, especially the $n = 6$ row and the actinides[8]. Because the solution of the fully relativistic Dirac equation is very complicated, the usual procedure is to add relativistic correction terms to the Schrödinger equation, Eq. (3.3). Relativistic corrections include the increase in electron mass with velocity and spin-orbit splitting. The mass-velocity effect results in the contraction and energetic stabilization of *s*- and *p*-orbitals and the consequent destabilization of *d*- and *f*-orbitals. Spin-orbit splitting leads to increased *s-p* hybridization energy. These effects have an important influence on the stable crystal structures of the heavy elements.

LDA total-energy calculations have been outstandingly successful in the accurate calculation of the ground-state structures of the solid elements. However, this method has important limitations, including the underestimation of band-gap energies and the failure to predict narrow-band Mott-transition phenomena. Typical LDA calculations are performed for a static lattice of nuclei. In this approach to the solid state, the zero-point phonon energy can be found by computing the total energy of the static lattice with unit cells distorted to correspond to various phonon wave vectors.

A more general method for computing the equilibrium state of matter at nonzero temperatures is the quantum-molecular-dynamics method of Car and Parrinello[9]. Here the LDA wave function is solved for a small number of nuclei in an arbitrary solid or liquid configuration, and the Hellman-Feynman theorem is used to find the net force on each nucleus. The nuclei are then moved in accord with Newtonian dynamics, and the LDA calculation is repeated for the new configuration of nuclei. This method has been useful in determining band structures and the details of bonding in solids and liquids at $T > 0$ K.

3.3 Interaction Potentials

Statistical-mechanical calculations of thermodynamic functions for solids and liquids require expressions for interatomic potentials. Although statistical theories can be extended to three- and higher n-body potentials, it is most convenient if the potentials are two-body and pairwise additive. These potentials unfortunately cannot be obtained directly from band-structure theory.

For simple metallic solids it is possible to expand the total energy about the uniform free-electron gas limit in powers of a weak electron-ion pseudopotential W[10,11]:

$$E_{tot} = E_K + E_{xc} + E_M + E^{(1)} + E^{(2)} + \dots \qquad (3.4)$$

The terms on the right hand side are electron-gas kinetic energy, electron-gas exchange and correlation energy, the Madelung (electrostatic) energy for point ions in a uniform electron gas, the correction to E_M due to the pseudopotential, and the band-structure energy due to the linear response of the electron gas to the pseudopotential. We assume Z_i valence electrons per ion.

The band-structure energy $E^{(2)}$ corresponds to an effective interionic potential in addition to the purely Coulombic repulsion contained in E_M[12]:

$$\phi(r) = \frac{Z_i^2 e^2}{r} + \frac{1}{2\pi^2}\int_0^\infty \phi_{BS}(k)\frac{\sin kr}{kr}k^2 dk\ , \qquad (3.5)$$

and

$$\phi_{BS}(k) = \frac{4\pi Z_i^2 e^2}{k^2}\left(\frac{1}{\varepsilon(k)} - 1\right)\left(\frac{k^2 W(k)}{4\pi Z_i e^2}\right)^2 . \qquad (3.6)$$

The first term in Eq. (3.5) is the bare-Coulomb repulsion between ions, and the second is the screening correction due to the modulations in electron density near the ions. The potentials arising from NFE perturbation theory have two interesting characteristics: 1) they are volume dependent because the dielectric function $\varepsilon(k)$ has volume dependence, and 2) they show an oscillatory structure, as shown in Fig. 3.1. Higher terms $E^{(3)}$, $E^{(4)}$, etc. in Eq. (3.4) correspond to 3-body, 4-body, etc., potential terms. Eqs. (3.5) and (3.6) can be extended easily to nonlocal pseudopotentials and may include exchange and correlation in the screening.

This nearly free-electron (NFE) perturbation theory works well for metals with plane-wave-like *sp* valence bands. When the more localized *d* and *f* electrons are important in bonding, however, then the simple perturbation theory breaks down. NFE pseudopotential theory has been extended to include localized orbitals in the Generalized Pseudopotential Theory (GPT)[13,14].

The GPT is a perturbative theory of the total energy of metals which applies to both nearly free and localized (non-free) bonding electrons.

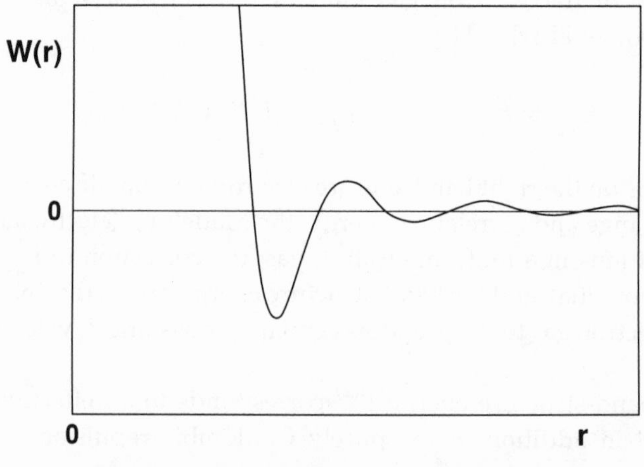

Fig. 3.1 A typical metallic pair potential.

Although this theory has been elaborated only for *d*-electron metals, it may also be applied to *f*-electron metals such as the actinides. The GPT begins with the same local-density equations used in band-structure theory, i.e., Eqs. (3.2) and (3.3) with the pseudopotential W defined with respect to the nonbonding ionic core. A mixed basis set of plane waves and localized atomiclike *d* states is introduced and simultaneous expansions are then developed for $n(r)$ and E_{tot} in terms of small quantities. These quantities are plane-wave matrix elements of the pseudopotential, which pertain to the *s* and *p* electrons (as in the NFE theory), and plane-wave *d* and interatomic *d*-*d* matrix elements, which pertain to the *d* electrons.

The GPT expansions become increasingly complex along the sequence: 1) NFE metal, 2) empty *d*-band metal, 3) filled *d*-band metal, and 4) partially filled *d*-band metal. As with the simple NFE model, the advantage of the GPT lies in the isolation of well-defined interatomic potential functions which can be used to compute the properties of any structure, solid or liquid, and thus to predict complete phase diagrams.

For insulators like the rare-gas solids, the NFE perturbation expansion is not valid, and another approach is needed. This may be done by QMC or *ab initio* molecular orbital calculations on small clusters of 2, 3, 4, etc. atoms or molecules[15,16]. For closed-shell atoms and molecules, these calculations reveal the form of the two-body interaction and the importance of the higher many-body contributions. Typical pair potentials have an attractive well and a steep repulsive core, as shown in Fig. 3.2. Although theoretical calculations can in principle generate intermolecular interactions, they are still not sufficiently accurate or convenient for widespread use. Instead, it is still common to fit effective pair potentials to experimental data. These may then be used directly in statistical theories of solids and liquids.

For diatomic and polyatomic molecules, the intermolecular potential function becomes very complex because of its angular dependence[17]. A useful procedure for fitting such potentials is the site-site potential, illustrated in Fig. 3.3. Here the total pair potential is represented as the sum of simple potentials centered on the nuclei or at other points in the molecules, plus electric multipole terms. The parameters in the potential are then fitted to thermodynamic data.

3.4 Statistical Models

Assuming purely classical nuclear motions, the solid and liquid states of particles with known interparticle potentials may be simulated with Monte Carlo (MC)[18] or Molecular Dynamics (MD)[19] methods. As with QMC,

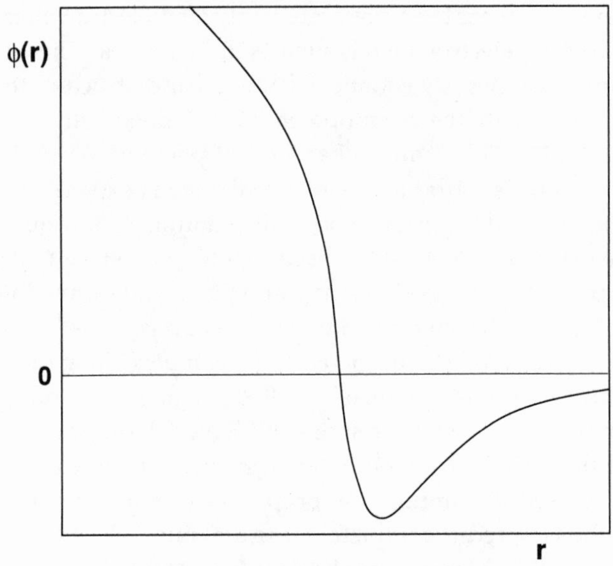

Fig. 3.2 A typical molecular pair potential.

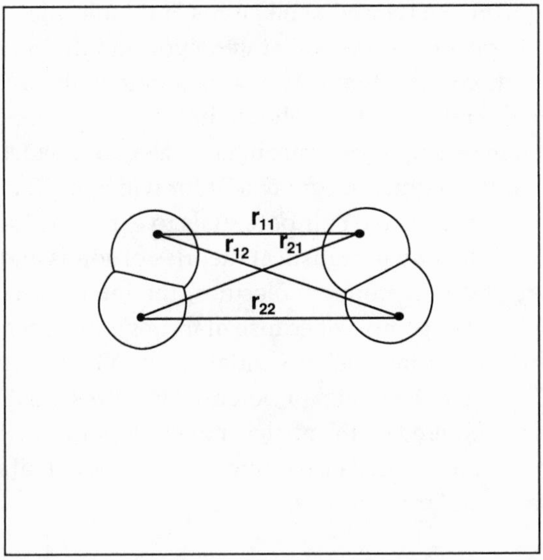

Fig. 3.3 The site-site interactions for a diatomic molecule.

these methods are in principle exact. In the MC method, an initial configuration of particles is placed in a finite box with periodic-boundary conditions and fixed temperature T. Random numbers are used to select and move individual particles. If the potential energy change ΔU resulting from a move is less than zero, then the new configuration is accepted; if $\Delta U > 0$, then it is accepted only with probability $\exp(-\Delta U/kT)$. This procedure converges to a state of local equilibrium at which accurate values of energy and pressure may be obtained. However, the small size of the system (typically a few hundred particles) prevents the occurrence of melting or solid-solid phase transitions at the equilibrium point. Also, although direct calculation of free energy is possible[20], it is not yet practical for determining phase stability. Thus while the MC method may be used to obtain the equation of state with good accuracy, it is not yet suitable for accurate mapping of phase boundaries.

In the MD method, Newton's equations of motion are solved for a few hundred particles in a box, and the pressure and energy may be computed at each time step. This system also converges to local thermodynamic equilibrium and has the same difficulty as MC in locating phase boundaries. Although the isobaric-isothermal ensemble versions of MC and MD have been used to predict the most stable crystal structures of certain solids[21], the most important use of these methods has been to provide a standard for comparing and refining approximate statistical-mechanical models.

For the solid state, the most useful model for computation of free energies is lattice dynamics[22,23]. Assume that the interatomic potential ϕ is pairwise additive and that the atoms move only small distances $\boldsymbol{u}(\boldsymbol{R})$ from their lattice sites at $\boldsymbol{R}$. Then the instantaneous potential energy is

$$U = \frac{1}{2}\sum_{\boldsymbol{RR'}} \phi(\boldsymbol{R} - \boldsymbol{R'} + \boldsymbol{u}(\boldsymbol{R}) - \boldsymbol{u}(\boldsymbol{R'})) \ . \tag{3.7}$$

Expanding this in a Taylor expansion yields

$$U = \frac{1}{2}\sum_{\boldsymbol{RR'}} \phi(\boldsymbol{R} - \boldsymbol{R'}) + \frac{1}{2}\sum_{\boldsymbol{RR'}} (\boldsymbol{u}(\boldsymbol{R}) - \boldsymbol{u}(\boldsymbol{R'})) \cdot \nabla\phi(\boldsymbol{R} - \boldsymbol{R'})$$

$$+ \frac{1}{4}\sum_{\boldsymbol{RR'}} [(\boldsymbol{u}(\boldsymbol{R}) - \boldsymbol{u}(\boldsymbol{R'})) \cdot \nabla]^2 \phi(\boldsymbol{R} - \boldsymbol{R'}) + \ldots \ . \tag{3.8}$$

The first term is the static-lattice energy. The second term vanishes because there is no net force on the atoms at their lattice positions. The third term is the harmonic energy, and in quasiharmonic lattice dynamics the series is truncated at this point. A plane-wave solution $\exp[i(\mathbf{k} \cdot \mathbf{R} - \omega t)]$ of the equations of motion for a harmonic lattice leads to the eigenvalue equation

$$M\omega^2\varepsilon = \mathbf{D}(\mathbf{k})\varepsilon, \tag{3.9}$$

where M is the particle mass, ω is the frequency, and $\mathbf{D}(\mathbf{k})$ is the dynamical matrix, related directly to the harmonic term in Eq. (3.8). This equation is solved for wave vectors $\mathbf{k}$ within the first Brillouin zone of the reciprocal lattice. For a lattice with an n-atom basis, there are $3n$ frequencies ω for each $\mathbf{k}$ value.

The free energy is then the sum over $3N$ thermally excited quantum harmonic oscillators:

$$A = U_o + \sum_{i=1}^{3N} \left(\frac{\hbar\omega_i}{2} + kT \ln[1 - \exp(-\frac{\hbar\omega_i}{kT})] \right), \tag{3.10}$$

where U_0 is the static-lattice potential energy.

For low temperatures and high pressures, the harmonic approximation is very good. However, for points near the melting curve, corrections for anharmonic motion are necessary for accurate free energies. Anharmonic corrections may be computed from higher terms in Eq. (3.8), but this is a rather slowly converging series[23]. In the self-consistent phonon model, corrections of all orders are obtained by self-consistently averaging the interaction potential over the finite amplitude of the atomic motion[23].

Liquid free energies are most efficiently calculated by variational theories[24]. These are based on the rigorous inequality

$$A \leq A_o + \frac{N}{2V} \int_0^\infty [\phi(r) - \phi_o(r)] g_o(r)\, 4\pi r^2 dr\ . \tag{3.11}$$

This means that the true Helmholtz free energy for a known interaction potential $\phi(r)$ is bounded above by the free energy for a simple reference potential $\phi_0(r)$ plus the potential difference averaged over the reference fluid configuration. What is needed for this calculation is the reference free energy A_0 and the reference-pair-distribution function $g_0(r)$ in analytic or tabular form. For each volume and temperature the right-hand side is evaluated for different values of the reference-fluid coupling parameter and the minimum is taken as the best approximation to A. The most readily available reference fluid is the hard-sphere fluid[24], but this yields a rather inaccurate A value. Much better results are obtained for smooth reference potentials, such as the $1/r^{12}$ potential[25].

There are many variations on these models of solids and liquids, but they have in common the attempt to achieve quantitative agreement with MC or MD calculations. This makes them suitable for accurate modeling of real solids and liquids.

3.5 Thermodynamics of Phase Transitions

A theoretical study of phase diagrams begins with the thermodynamics of first-order phase transitions. For two phases 1 and 2, the condition for thermodynamic equilibrium is the equality of temperature, pressure, and Gibbs free energy, $T_1 = T_2$, $P_1 = P_2$, and $G_1 = G_2$. The principal constraint on these transitions is the Phase Rule[26]:

$$f = c - p + 2. \tag{3.12}$$

Here f is the number of degrees of freedom in the temperature-pressure-composition space, c is the number of components, and p is the number of phases. Elements under the pressure and temperature conditions of interest are one-component systems, so $c = 1$ and $f = 3 - p$. Hence a single phase ($p = 1$) is represented by an area ($f = 2$) in the P-T plane, a two-phase mixture ($p = 2$) is represented by a curve ($f = 1$) (phase line or boundary), and a three-phase mixture ($p = 3$) is represented by a point ($f = 0$) (triple point). A one-component phase diagram is a plot of areas representing phases, which are demarcated by phase boundaries in the P-T plane.

A typical one-component phase diagram will exhibit a gas or vapor phase, a liquid phase, and several solid phases, as shown schematically in Fig. 3.4. The liquid-gas boundary ends in a critical point. Except in the case of helium, the solid-gas, liquid-gas, and solid-liquid phase boundaries intersect at a triple point. The liquid-vapor transition is discussed separately in chapter 16.

A few useful thermodynamic rules concerning phase boundaries are available. The Clausius-Clapeyron equation relates the entropy and volume changes across a phase boundary to the slope of the boundary:

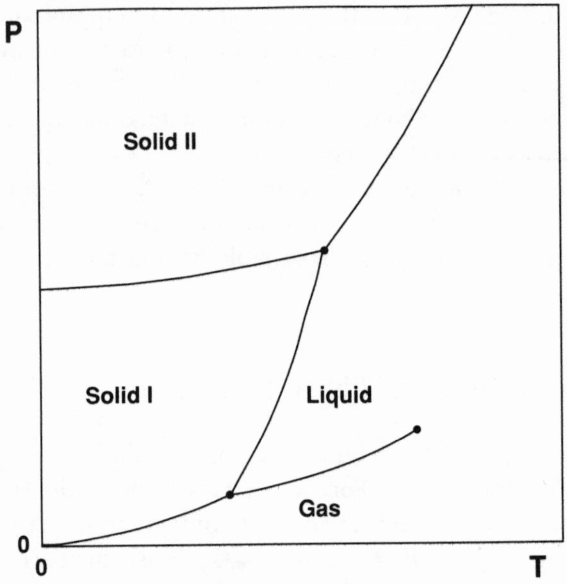

Fig. 3.4 A schematic phase diagram showing solid-gas, liquid-gas, solid-liquid, and solid-solid phase boundaries.

$$\frac{dP}{dT} = \frac{\Delta S}{\Delta V}. \tag{3.13}$$

The Second Law of thermodynamics requires $\Delta S > 0$ for increasing temperature, but ΔV can be of either sign. Hence dP/dT can be either positive or negative. Melting curves usually show a positive slope, but solid-solid transitions do not show a strong preference in sign.

The second derivative d^2P/dT^2 can be written[27]

$$\frac{d^2P}{dT^2} = -\frac{1}{\Delta V}\left[\frac{d\Delta V}{dP}\left(\frac{dP}{dT}\right)^2 + 2\,\frac{d\Delta V}{dT}\frac{dP}{dT} - \frac{\Delta C_P}{T}\right]. \tag{3.14}$$

For low-compressibility materials, the terms on the right-hand side are small, so that the phase boundaries will have very small curvature and hence will look like straight lines over the experimentally available pressure ranges.

The Third Law of thermodynamics requires that any phase boundary which touches the 0 K axis must have zero slope, $dP/dT = 0$.

Practical calculation of phase diagrams requires the location of all phase transitions over a specified range of T and P. This is usually done by plotting isotherms of A vs. V or G vs. P for each possible phase. For the A-V isotherms, a phase transition is indicated by a common tangent making contact with two A curves. The two volumes at the contact points are the equilibrium volumes, and the negative slope of the tangent line is the equilibrium pressure. For the G-P isotherms, the intersection of two curves locates the equilibrium pressure, and the slopes of the two curves at the intersection point are the equilibrium volumes. These plots are illustrated in Fig. 3.5. A series of such plots at different T values is required to map out a complete phase boundary.

3.6 Melting

The object of theories of melting is to characterize accurately the thermodynamic properties along the melting curve[28,29]. There are two important

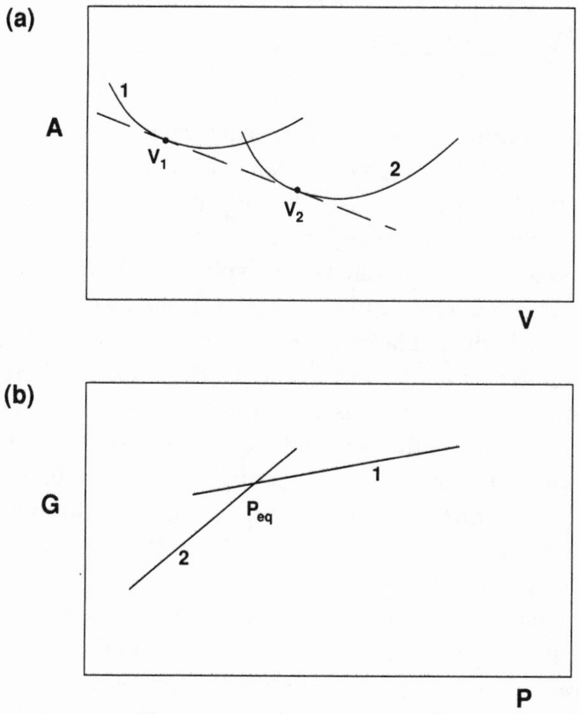

Fig. 3.5 Method of calculating phase transitions from theoretical free-energy functions. In (a), the equilibrium volumes are determined by a common tangent between Helmholtz free energy isotherms of phases 1 and 2. In (b), the equilibrium pressure is determined by the intersection of the two Gibbs free energy isotherms.

semiempirical "one-phase" theories which predict melting on the basis of the properties of either solid or liquid alone. The first of these, the Lindemann Law, is based on the hypothesis that the root-mean-square atomic displacement in the solid is a constant fraction of the interatomic spacing a at the melting point[30]:

$$\left(\frac{<R^2>}{a^2}\right)^{1/2} \approx 0.1 \; . \tag{3.15}$$

Use of quasiharmonic lattice dynamics recasts this law into the form

$$\frac{T_m}{M{\Theta_D}^2 V^{2/3}} = \text{const} \; , \tag{3.16}$$

or, equivalently,

$$-\frac{d\,\ln T_m}{d\,\ln V} = 2\gamma(V) - 2/3 \, , \tag{3.17}$$

where T_m is the melting temperature, Θ_D is the Debye temperature, M is the atomic mass, and $\gamma(V)$ is the Grüneisen parameter. Together with an equation of state $P(V,T)$, this rule can be used to compute a melting curve.

Another one-phase melting rule is due to Ashcroft and Lekner[31]. These authors noticed that if the theoretical hard-sphere fluid structure factor $S(k)$ is fitted to the experimental liquid $S(k)$ at the freezing point, the best fit is obtained for a hard-sphere packing fraction $\eta = \pi N\sigma^3/6V \cong 0.45$, where σ is the sphere diameter. As indicated by the results of hard-sphere perturbation theory, this relation also holds along the melting curve as a function of pressure[25]. In practice, this means that the freezing point is predicted wherever the perturbation theory yields $\eta \cong 0.45$. Both the Lindemann and Ashcroft-Lekner melting rules are capable of semiquantitative agreement with experiment and they are also capable of representing anomalous melting curves where $dP/dT < 0$.

A thermodynamically rigorous theory of melting requires the free energies of both phases. A good way to test melting models is to carry out accurate theoretical calculations with simple model systems. Here we assume that the particles interact according to a pairwise-additive potential function, and we then proceed by various means to compute the thermodynamic properties and the melting curve.

A very simple class of potentials with only one characteristic parameter are the repulsive inverse-power potentials:

$$\phi(r) = \varepsilon\left(\frac{\sigma}{r}\right)^n . \tag{3.18}$$

The case $n = \infty$, or the hard-sphere potential, has been extensively studied by both MC and MD computer simulations[32,33]. The "soft-sphere" potentials $n = 12, 9, 6,$ and 4 have also been accurately characterized[34]. For $n \leq 3$, the energy of the system is divergent, but it can be made finite by adding a uniform compensating background of opposite "charge." The attraction between the particles and the background then cancels the infinity in the repulsive energy. For $n = 1$, this is the one-component plasma (OCP)[35].

Hard spheres are characterized by the sphere diameter σ, or the closest-packing volume $V_0 = N\sigma^3/\sqrt{2}$. Hence a useful coupling parameter is the dimensionless ratio V/V_0. Computer simulations of this system reveal a transition from a close-packed solid phase to a liquid phase[32,33]. The solidus and liquidus volumes are $(V/V_0)_s = 1.35$ and $(V/V_0)_l = 1.50$. The reduced volume and entropy changes upon melt are $\Delta V/V_s = 0.11$ and $\Delta S/Nk = 1.20$, which accord well with measurements on the monatomic elements. Thus the extremely simple hard-sphere system already contains the necessary ingredients for an accurate theory of melting. It is interesting that for hard spheres at the melting point $(\langle R^2\rangle/a^2)^{1/2} = 0.125$ and $\eta = 0.49$, both constant along the melting curve and in semiquantitative agreement with the one-phase models of melting[36].

For $3 < n < \infty$, an appropriate coupling parameter is $z = (N\sigma^3/\sqrt{2}V)(\varepsilon/kT)^{3/n}$. Simulations have been made for $n = 12, 9, 6,$ and 4[34]. Each of these potentials shows a melting transition, again with experimentally realistic characteristics. In general, the form of the P-T melting curve is $P = \text{const} \times T^{1+3/n}$.

For the OCP with $n = 1$ and with a uniform neutralizing background, the coupling parameter is $\Gamma = Z^2e^2/kTR_A$, with $R_A = (3V/4\pi N)^{1/3}$. Again, extensive computer simulations have revealed solid and liquid phases separated by a melting curve[35]. The OCP melts at $\Gamma \approx 178$, with $\Delta S/Nk \approx 0.8$. Because of the uniform background, it happens that $\Delta V = 0$[37]. The melting characteristics for the inverse-power potentials are given in Table 3.2, and the melting curves are shown in Fig. 3.6.

The entropy and volume changes at melting predicted by the simple one-parameter models are in good agreement with experimental data on molecular and metallic elements. Also, the measured exponents in the empirical Simon law, $P = a + bT^c$, are mostly within the range predicted by the inverse-power scaling relation[28,29]. These results suggest that simple repulsive potentials realistically mimic melting.

TABLE 3.2 Melting Properties of the Inverse-Power Potentials

n	1	4	6	9	12	∞
z_s	—	3.94	1.56	0.971	0.844	0.736
z_l	—	3.92	1.54	0.943	0.813	0.667
Γ	178	—	—	—	—	—
$\Delta S/R$	0.8	0.80	0.75	0.84	0.90	1.20
$\Delta V/V_s$	0	0.005	0.013	0.030	0.038	0.11
$(<R^2>/a^2)^{1/2}$	0.1	0.14	0.15	0.15	0.15	0.125

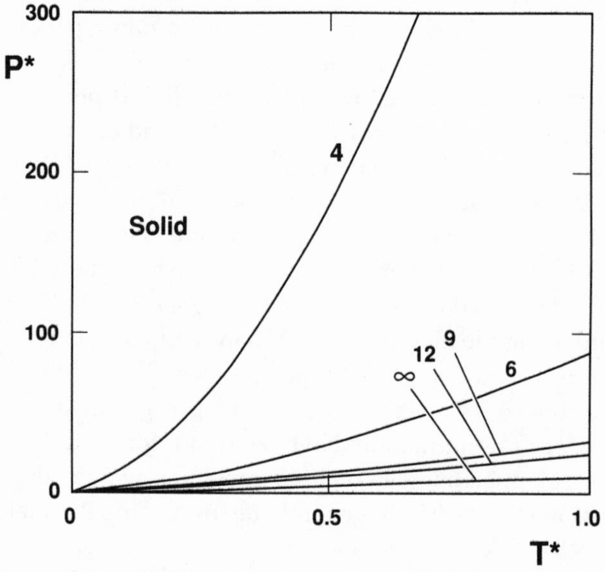

Fig. 3.6 The melting curves of the inverse-power potentials.

Smooth monotonic melting curves can be modeled readily with the one-parameter potentials. More complex melting curves require more complex potentials. The addition of an attractive term to a repulsive potential leads to a gas phase, a triple point, and a liquid-gas phase boundary which ends in a critical point. The phase diagram of the Lennard-Jones 6–12 potential has been worked out in detail by computer simulation[38], and is shown in Fig. 3.7. For general potentials like the 6–12, the melting curve can be calculated from statistical-mechanical theory. This involves a quasiharmonic lattice dynamics calculation together with an anharmonic correction for the solid free energy and a variational calculation for the fluid free energy. With pair potentials fitted to experimental data, these models are now

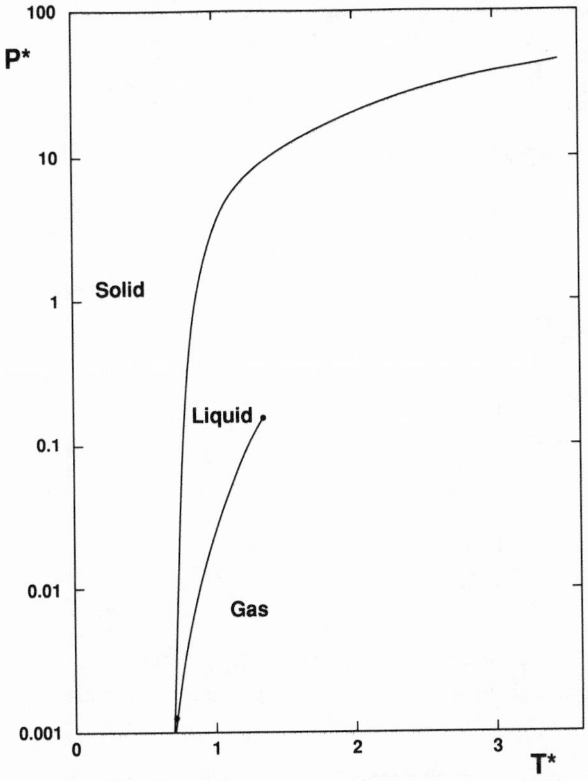

Fig. 3.7 A semilogarithmic plot of the phase diagram of the Lennard-Jones 6–12 potential. The melting curve and liquid-gas boundaries are shown.

sufficiently accurate that quantitative agreement between experiment and theory is possible[39]. These calculations are most readily applied to molecular and metallic phases where the pair-potential concept is considered valid.

Melting curves showing temperature maxima and minima require inflections which "soften" the potential in narrow ranges of volume. This softening causes a lowering of the melting temperature and thus a negative value of dP/dT. It may also cause a solid-solid transition with a large volume change. These unusual effects, which mimic electronic phase transitions, have been verified by computer simulations of a two-dimensional system of particles with a repulsive "step" potential[40], as shown in Fig. 3.8.

The high-pressure limit of the melting curve has been of interest since the early years of this century[41]. Three possible curves are sketched in Fig. 3.9. The melting curve might 1) end in a critical point, 2) continue

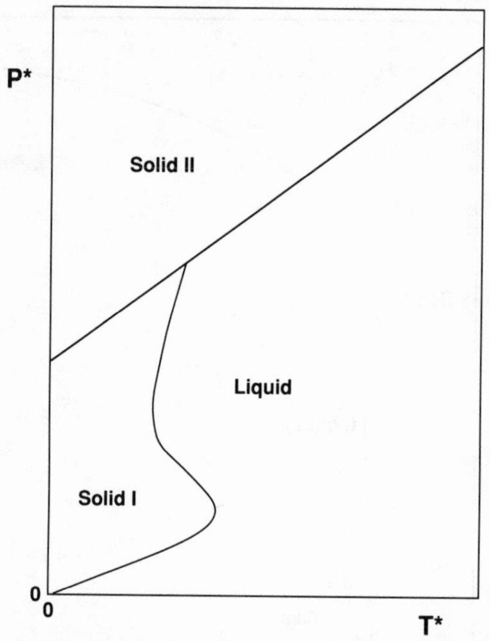

Fig. 3.8 The step-potential phase diagram in two dimensions. A maximum in the melting curve and a solid-solid phase transition are found.

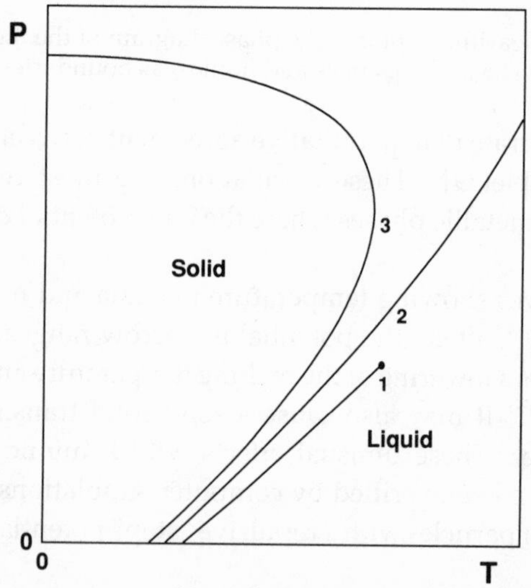

Fig. 3.9 The possible fates of the melting curve. The curve may (1) end in a critical point, (2) increase in temperature and pressure indefinitely, or (3) form a closed loop.

increasing monotonically, or 3) end at a maximum pressure at $T = 0$ K. It is now clear that option 3) is correct[42].

In the electron-gas model commonly used in solid-state physics, there is a phase transition upon compression from a solid to a fluid phase at $r_s = (3V/4\pi NZ)^{1/3} = 100 \pm 20$ bohr, as determined by QMC[43]. This transition occurs because of quantum delocalization of the electrons, since the electron vibrational energy increases with electron density as $\rho^{1/2}$ while the binding energy increases only as $\rho^{1/3}$.

At $T = 0$ K and enormous pressures, solid matter is composed of bare nuclei in a nearly uniform electron gas. This is the "inverse" of the electron lattice but the physics of the 0 K melting will be identical with the electron case, except for the mass difference and the possibility that the nuclei may be bosons rather than fermions[44]. The existence of 0 K melting means that the melting curve forms a closed loop and that there exists a maximum melting temperature, as shown in Fig. 3.9, curve 3.

In summary, it looks as though our current understanding of melting in the monatomic elements is good. Improvements primarily in quantitative accuracy are still needed. More complex molecular and covalent solid and liquid states are still not adequately dealt with because of the difficult problem of computing free energies.

3.7 Solid-Solid Transitions

As with melting, the calculation of solid-solid transitions requires the free energies of the two phases in equilibrium. This is generally a more difficult theoretical problem than melting because the very small volume and energy changes across a typical solid-solid transition mean that very high accuracy in the free-energy functions is required for adequate prediction of the transition.

It is useful to examine simple potentials in order to understand systematic trends in crystal structures. Accurate computer simulations and theoretical work on hard spheres indicates that the stable crystal structure is probably fcc[20,45]. For inverse-power potentials, the most stable lattice at high density is also fcc, but for $n < 7.6$, a bcc phase appears before melt[46]. This is shown in Fig. 3.10. Although the fcc structure has a lower internal energy for inverse-power potentials, the bcc structure has a higher entropy, and is therefore stabilized at high temperatures. As n decreases, the bcc phase occupies an increasing fraction of the phase diagram. The OCP solid phase is entirely bcc. These facts immediately explain why so many elements have the fcc, hcp, bcc, or closely related structures, and they also suggest that the

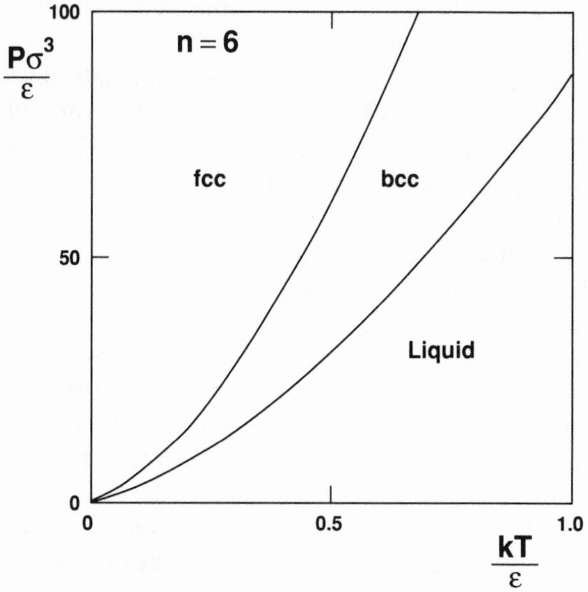

Fig. 3.10 The fcc-bcc equilibrium for the inverse-6th-power potential.

"softness" of the interatomic repulsions plays a role in the very common temperature-induced close-packed-to-bcc phase transitions found in many elements.

For NFE metals, it is the oscillating behavior in the pair potential arising from the electron-ion interaction together with the volume-dependent mean-field energy that allows for a wide range of crystal structures. By varying the valence Z_i, the pseudopotential core radius r_c, and the electron radius r_s in the NFE pseudopotential model, variations in the effective pair potential occur which make the spectrum of structures of the nontransition elements understandable[47–49]. When the attractive potential well occurs at the nearest-neighbor distance, a close-packed or bcc structure will be stable. The alignment of the oscillations with the neighbor shells will determine which structure will be the stable phase. However, the competition between the pair potential and the mean field can lead to a local maximum in the potential at the nearest-neighbor distance. In this case, the crystal lattice will undergo a distortion so that the energy can be lowered. This gives rise to lower-symmetry structures in such elements as Ga and Hg. Although covalent structures cannot be predicted by this technique, it does strongly indicate the stability for open lattices such as simple cubic for Group V elements at high pressure. The NFE model also illustrates the trend of structure changes under compression, and can be used in lattice-dynamics and liquid calculations to generate complete phase diagrams.

For the transition metals, the GPT predicts large 3-body potential terms which make an important contribution to the stability of the observed fcc, hcp, and bcc structures[50].

For molecular solids such as N_2, the usual approach to structure calculations is to fit an angle-dependent pair potential to experimental equation of state data and then to use this potential to find the lowest energy structures by a trial-and-error method[51]. There is as yet no general theory of diatomic structures at high pressure.

For a completely general study of solid-solid transitions at 0 K, total-energy band-structure calculations are done[6,7]. This has been the prime source of phase-transition predictions for the covalent elements. Band-structure theory can also be used to predict metallization of an insulator even when no first-order phase transition is indicated.

3.8 Liquid-Liquid Transitions

In two-component systems, first-order phase transitions between two dense fluid phases are common[52]. A mixture of components A and B can separate into A-rich and B-rich coexisting phases. In one-component systems, however, liquid-liquid transitions are exceedingly rare. Among the elements, the only confirmed example is found in He-3, and this is a phenomenon of quantum statistics. Phase-transition-like changes are found in liquid sulfur and are probably due to equilibria among various polymeric species.

No thermodynamic principle is violated by liquid-liquid transitions in one-component systems, and their rarity must have to do with the nature of the liquid state. That the liquid has no long-range order precludes small volume changes such as are seen in the solid between two close-packed phases. The liquid smoothly changes its density through the same region where a density discontinuity occurs in the solid. Even in such large-volume solid-solid transitions as in Cs or Ce, there is no evidence of any discontinuity in the liquid.

Theoretical models of the dissociation of diatomic fluids at high pressure and temperature have led to predictions of first-order transitions ending in a critical point[53], but there is as yet no experimental confirmation of this. Also, there have been published claims for the existence of phase transitions in the plasma state far from the melting curve, but so far the experimental evidence is not conclusive[54]. If such phase transitions are ever confirmed, they would become a legitimate part of phase-diagram studies, and their systematics would be of great theoretical interest.

3.9 Generalized Phase Diagrams

One of the main theoretical problems of phase-diagram studies is to explain the common patterns seen in columns or rows of elements. Frequently the same sequence of phases is found in a series of related elements, but with different characteristic temperatures and pressures.

For a simple one-parameter potential of the kind described previously, a phase transition is characterized by a constant value of that parameter. In the *P-T* plane this represents a smooth curve passing through the origin, $T = 0, P = 0$. For a two-parameter potential, for example a Lennard-Jones 6–12 potential, a universal phase diagram can be constructed in reduced *P-T* space, as shown in Fig. 3.7. The reduced variables are $P^* = P\sigma^3/\varepsilon$ and $T^* = kT/\varepsilon$, where σ and ε are the characteristic distance and energy for the potential. This means that any two phase diagrams arising from this potential with different values of σ and ε can be related to each other by multiplicative coefficients: $P_2 = aP_1, T_2 = bT_1$. The rare-gas phase diagrams are a close approximation to this "corresponding-states" principle.

For potentials with three or more variable parameters, the phase diagrams cannot each be reduced to the same universal diagram. For metals in the pseudopotential theory, there are three characteristic length scales[49], the pseudopotential core radius r_c, the electron screening length λ_{sc}, and the Friedel wavelength $2\pi/2k_F$. In passing from one element to a related one, all of these parameters change, leading to a complex pattern of phase changes. To a rough first approximation, we can think of the total energy as divided into pair-potential and mean-field parts. Variation of the mean field leads to an approximately constant pressure shift, while variation of the pair potential leads to multiplicative shifts in temperature and pressure. Hence the scaling becomes $T_2 = aT_1$, $P_2 = bP_1 + c$. This implies the existence of a "global" diagram for the group of elements of which each element represents only a part. It is this "quasi-periodic" behavior of the elements that challenges the theory.

3.10 Extreme Conditions

Matter at low temperature and pressure has its most complex structure because here the outer electrons are only slightly perturbed and can manifest themselves in complex types of bonding and crystal structure. Application of high temperature or high pressure begins to reduce the differences among the elements by delocalizing the bonding electrons. This process of "smoothing" of atomic properties is well advanced by $T = 10$ eV or $P = 10$ TPa[42]. Under these conditions, thermodynamic functions and phase diagrams become more predictable theoretically.

At still higher temperatures and pressures, the nuclei begin to undergo reactions and the system can no longer be considered a pure chemical element[42]. In hydrogen this reaction is the fusion process, and it begins at $T = 100 - 1000$ eV (thermonuclear) or $\rho = 10^4 - 10^5$ Mg/m^3 (pycnonuclear). For a stable nuclide such as Fe the reaction is dissociation at high temperatures ($T = 10^5 - 10^6$ eV) or neutronization at high pressures ($P = 10^6$ TPa). These extreme pressure and temperature points join to form smooth curves in P-T space which delimit the area considered in this book.

3.11 Discussion

The theories described in this chapter have been remarkably successful in predicting experimentally measured phase transitions. This results both from improvements in theory and a steady increase in computer power. It is now possible to perform accurate computer simulations of many-body systems, then to refine approximate theories to fit the simulations, and finally to apply the theory to experimental measurements. Insight and experience are gained from computer "experiments" on many-particle systems with simple interactions without reference to physical experiment. Powerful computers then allow an iterative process of improving theoretical models until accurate agreement with simulation is obtained. The result has been a set of accurate theories which can be applied to real materials with confidence.

The growth of theory has occurred simultaneously with the advent of the DAC, and the result has been an accelerating cycle of calculations and measurements which have enriched condensed-matter physics. With the continued growth of computer power, we may expect further rapid theoretical progress in the years ahead.

For Further Reading

N. W. Ashcroft and N. D. Mermin, *Solid State Physics* (Saunders College, Philadelphia, 1976).

J. Ihm, "Total Energy Calculations in Solid State Physics," Rep. Prog. Phys. **51**, 105 (1988).

H. L. Skriver, *The LMTO Method* (Springer, Berlin, 1984).
J. Hafner, *From Hamiltonians to Phase Diagrams* (Springer, Berlin, 1987).
W. A. Harrison, *Electronic Structure and the Properties of Solids* (Freeman, San Francisco, 1980).
D. C. Wallace, *Thermodynamics of Crystals* (Wiley, New York, 1972).
J. P. Hansen and I. R. McDonald, *Theory of Simple Liquids*, 2d ed. (Academic, London, 1986).
H. B. Callen, *Thermodynamics* (Wiley, New York, 1963).
J. E. Ricci, *The Phase Rule and Heterogeneous Equilibrium* (Dover, New York, 1966) (reprint of 1951 edition).
W. G. Hoover and M. Ross, "Statistical Theories of Melting," Contemp. Phys. **12**, 339 (1971).
S. M. Stishov, "The Thermodynamics of Melting of Simple Substances," Usp. Fiz. Nauk **114**, 3 (1974) [Sov. Phys. Usp. **17**, 625 (1975)].
D. A. Kirzhnits, "Extremal States of Matter (Ultrahigh Pressures and Temperatures)," Usp. Fiz. Nauk **104**, 489 (1971) [Sov. Phys. Usp. **14**, 512 (1972)].

References

1. D. Ceperley and B. Alder, Science **231**, 555 (1986).
2. N. W. Ashcroft and N. D. Mermin, *Solid State Physics* (Saunders College, Philadelphia, 1976) chaps. 8–11.
3. W. Kohn and P. Vashishta, in *Theory of the Inhomogeneous Electron Gas*, S. Lundqvist and N. H. March, eds. (Plenum, New York, 1983) chap. 2.
4. J. C. Slater, Phys. Rev. **81**, 385 (1951).
5. L. Hedin and B. I. Lundqvist, J. Phys. C **4**, 2064 (1971).
6. H. L. Skriver, *The LMTO Method* (Springer, Berlin, 1984).
7. J. Ihm, Rep. Prog. Phys. **51**, 105 (1988).
8. P. Pyykkö, Chem. Rev. **88**, 563 (1988).
9. R. Car and M. Parrinello, in *Simple Molecular Systems at Very High Density*, A. Polian, P. Loubeyre, and N. Boccara, eds. (Plenum, New York, 1989) p. 455.
10. W. A. Harrison, *Pseudopotentials in the Theory of Metals* (Benjamin, Reading, Mass., 1966).
11. J. Hafner, *From Hamiltonians to Phase Diagrams* (Springer, Berlin, 1987).
12. N. W. Ashcroft and D. Stroud, Solid State Phys. **33**, 1 (1978).
13. J. A. Moriarty, Int. J. Quantum Chem. Symp. **17**, 541 (1983).
14. J. A. Moriarty, Phys. Rev. B **38**, 3199 (1988).
15. D. M. Ceperley and H. Partridge, J. Chem. Phys. **84**, 820 (1986).
16. F. H. Ree and C. F. Bender, Phys. Rev. Lett. **32**, 85 (1974).
17. M. S. H. Ling and M. Rigby, Mol. Phys. **51**, 855 (1984).
18. K. Binder, ed., *Monte Carlo Methods in Statistical Physics*, 2d ed. (Springer, Berlin, 1986).
19. W. G. Hoover, *Molecular Dynamics* (Springer, Berlin, 1986).
20. D. Frenkel and A. J. C. Ladd, J. Chem. Phys. **81**, 3188 (1984).
21. M. Parrinello and A. Rahman, J. Appl. Phys. **52**, 7182 (1981).
22. Ref. 2, chap. 22.

23. D. C. Wallace, *Thermodynamics of Crystals* (Wiley, New York, 1972).
24. J. P. Hansen and I. R. McDonald, *Theory of Simple Liquids,* 2d ed. (Academic, London, 1986) chap. 6.
25. D. A. Young and F. J. Rogers, J. Chem. Phys. **81**, 2789 (1984).
26. H. B. Callen, *Thermodynamics* (Wiley, New York, 1963) chap. 9.
27. J. R. Partington, *An Advanced Treatise on Physical Chemistry* (Longmans, London, 1957) vol. III, pp. 494–495.
28. W. G. Hoover and M. Ross, Contemp. Phys. **12**, 339 (1971).
29. S. M. Stishov, Usp. Fiz. Nauk **114**, 3 (1974) [Sov. Phys. Usp. **17**, 625 (1975)].
30. V. N. Zharkov and V. A. Kalinin, *Equation of State of Solids at High Pressures and Temperatures* (Consultants Bureau, New York, 1971) pp. 75–79.
31. N. W. Ashcroft and J. Lekner, Phys. Rev. **145**, 83 (1966).
32. W. G. Hoover and F. H. Ree, J. Chem. Phys. **49**, 3609 (1968).
33. B. J. Alder, W. G. Hoover, and D. A. Young, J. Chem. Phys. **49**, 3688 (1968).
34. W. G. Hoover, S. G. Gray, and K. W. Johnson, J. Chem. Phys. **55**, 1128 (1971).
35. W. L. Slattery, G. D. Doolen, and H. E. DeWitt, Phys. Rev. A **26**, 2255 (1982).
36. D. A. Young and B. J. Alder, J. Chem. Phys. **60**, 1254 (1974).
37. J. D. Weeks, Phys. Rev. B **24**, 1530 (1981).
38. J. P. Hansen and L. Verlet, Phys. Rev. **184**, 151 (1969).
39. D. A. Young and M. Ross, J. Chem. Phys. **74**, 6950 (1981).
40. D. A. Young and B. J. Alder, J. Chem. Phys. **70**, 473 (1979).
41. Ref. 27, pp. 487–492.
42. D. A. Kirzhnits, Usp. Fiz. Nauk **104**, 489 (1971) [Sov. Phys. Usp. **14**, 512 (1972)].
43. D. M. Ceperley and B. J. Alder, Phys. Rev. Lett. **45**, 566 (1980).
44. R. Mochkovitch and J. P. Hansen, Phys. Lett. **73A**, 35 (1979).
45. K. W. Kratky, Chem. Phys. **57**, 167 (1981).
46. W. G. Hoover, D. A. Young, and R. Grover, J. Chem. Phys. **56**, 2207 (1972).
47. V. Heine and D. Weaire, Solid State Phys. **24**, 249 (1970).
48. C. W. Krause and J. W. Morris, Jr., Acta Met. **22**, 767 (1974).
49. J. Hafner and V. Heine, J. Phys. F **13**, 2479 (1983).
50. J. A. Moriarty, Phys. Rev. Lett. **55**, 1502 (1985).
51. V. Chandrasekharan, R. D. Etters, and K. Kobashi, Phys. Rev. B **28**, 1095 (1983).
52. J. S. Rowlinson and F. L. Swinton, *Liquids and Liquid Mixtures,* 3d ed. (Butterworths, London, 1982).
53. D. C. Hamilton and F. H. Ree, J. Chem. Phys. **90**, 4972 (1989).
54. V. A. Alekseev, V. E. Fortov, and I. T. Yabukov, Usp. Fiz. Nauk **139**, 193 (1983) [Sov. Phys. Usp. **26**, 99 (1983)].

CHAPTER 4
Hydrogen

4.1 Introduction

The simplicity of the hydrogen molecule and its large quantum effects in the solid state have led to a vast and growing experimental and theoretical literature on condensed hydrogen. The importance of hydrogen in astrophysics, the proposed use of cryogenic hydrogen-isotope fuels in fusion reactors, and the prediction of an exotic metallic-hydrogen phase have all contributed to this interest. With the advent of the DAC, experimental work on hydrogen under pressure has accelerated sharply. There is at present an intense and fruitful interplay between experiment and theory in the study of the phase diagrams of the hydrogen isotopes.

4.2 The Low-Pressure Solid

At low temperatures and pressures, hydrogen is a liquid or solid composed of diatomic molecules. Because the electron-density distribution around the nuclei is nearly spherical, the molecule is very nearly a free rotor even in the solid phase. However, because of the small nuclear mass, the rotational energies are widely spaced, and the selection of rotational states permitted by symmetry considerations in the homonuclear molecules H_2, D_2, and T_2 has a significant effect on the phase diagram[1]. Since the protons in H_2 are fermions, the wave function must be antisymmetric upon exchange of nuclei. The relevant part of the total wave function is the product of the nuclear-spin and rotational functions. Hence when the nuclear spins are in the $I = 0$ or $I = 2$ (antisymmetric) state, the allowed rotational functions are symmetric, or $J = 0, 2, 4,\ldots$. When the nuclear spins have the $I = 1$ (symmetric) state, the allowed rotational functions are antisymmetric, with $J = 1, 3, 5,\ldots$. In the case of deuterium, the nuclei are bosons and the product of the nuclear-spin and rotational wave functions must be symmetric upon exchange of nuclei. The nuclear-spin configuration with the lowest multiplicity is designated para, and with the highest, ortho. Although the thermodynamic equilibrium state of H_2 or D_2 will always be a mixture of $J = 0, 1, \ldots$ states, it is possible to prepare nearly pure $J = 0$ and $J = 1$ condensed phases because the transition between them is strongly forbidden. These facts are summarized in Table 4.1. The heteronuclear molecules HD, HT, and DT do not have indistinguishable nuclei, and hence all rotational states are permitted.

Experimental crystallography is difficult in solid hydrogen because of its transparency to X rays, and there has been much controversy in the past about the correct crystal structure of the solid hydrogens[1,2]. However, there is now wide agreement that $J = 0$ H_2 and D_2 are hex(2) (hcp), with the molecules freely rotating and their centers located on the lattice sites[1]. Because of the radioactivity of T_2, very few experimental data have been accumulated on this isotope[1]. Recent crystallographic work[3–5] suggests that the low-pressure hcp lattice has very nearly ideal packing, i.e., $c/a = (8/3)^{1/2}$.

Solidification of the metastable long-lived $J = 1$ molecules produces a low-temperature cubic sc(4) (Pa3) phase in which the rotational axes of the molecules orient along body diagonals of the unit cell[1]. This structure is shown in Fig. 4.1. Because the $J = 1$ molecules have asymmetric p-like wave functions, they can interact through the molecular electric quadrupole (EQQ) interaction, and theoretical static-lattice calculations show that Pa3 is in fact the preferred configuration for quadrupoles[1]. As the Pa3 phase is heated, it passes through a first-order transition to the hcp phase at a temperature determined by the $J = 1$ concentration. By cycling through the transition temperature, it is possible to separate the rotational order-disorder transition from the change in crystal symmetry. That is, one can

TABLE 4.1 Allowed Rotational States in the Hydrogen Isotopes

Isotope	Nuclear statistics	Overall wave function symmetry	Nuclear spin	Allowed rotational states	Name
H_2	fermion	antisymmetric	$I = 0$	$J = 0,2,4,\ldots$	para
			$I = 1$	$J = 1,3,5,\ldots$	ortho
D_2	boson	symmetric	$I = 1$	$J = 1,3,5,\ldots$	para
			$I = 0,2$	$J = 0,2,4,\ldots$	ortho
T_2	fermion	antisymmetric	$I = 0$	$J = 0,2,4,\ldots$	para
			$I = 1$	$J = 1,3,5,\ldots$	ortho

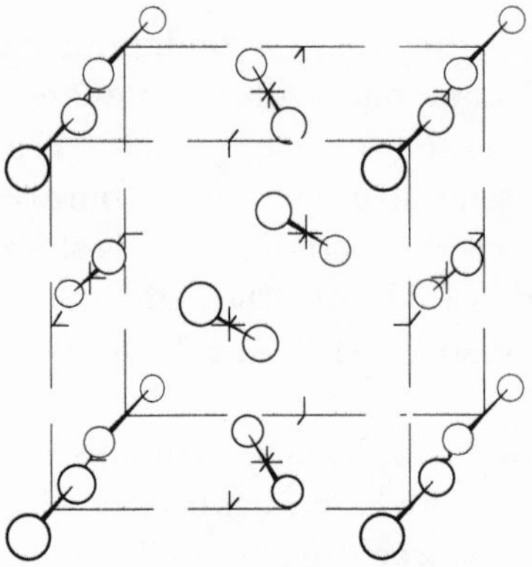

Fig. 4.1 The Pa3 crystal structure. (From T. A. Scott, Phys. Repts. **27**, 89 [1976]. Redrawn with permission.)

pass from an ordered Pa3 sc(4) to a disordered fcc(4) structure via a λ-transition. A plot of the transition temperature vs. the $J = 1$ concentration is shown in Fig. 4.2. Because of the very small difference in free energies between the Pa3 and hcp phases, it has not yet been possible to predict the location of the transition between them when lattice motions are taken into account. At very low $J = 1$ concentrations, a "quadrupolar glass" of randomly oriented rotational states is found. The Pa3-hcp phase line has been followed in $J = 1$ H_2 and D_2 up to about 0.6 GPa, and is shown in Fig. 4.3. The transition temperature increases with pressure[1].

4.3 High-Pressure Rotational Transition and Melting

Compression of the $J = 0$ molecular solids to very high pressures is expected to lead to a rotationally ordered phase because the anisotropic crystal-field potential, consisting of EQQ and repulsive valence-electron components, becomes stronger with increasing density and perturbs the free-rotor wave functions. The perturbations lead to the mixing of the $J = 0$ and higher J values, which are spatially asymmetric. These asymmetric rotational states are expected to "lock in" to an ordered librating state at a critical density.

Raman spectroscopy on H_2 and D_2 compressed in a DAC shows evidence of the predicted phase transition[6,7]. As the solid is compressed, a branch of the phonon spectrum approaches and then crosses the roton bands. This is followed at higher compression by a decrease in frequency (softening) of the rotons and finally by a strong broadening and splitting of the roton bands as the ordering occurs. In D_2 this occurs at 0 K and 27.8 GPa, with the transition pressure increasing with temperature[6]. The rotational

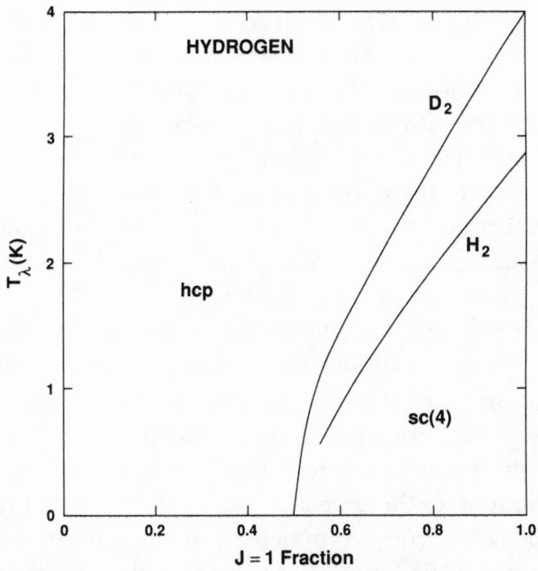

Fig. 4.2 The order-disorder transition temperature as a function of $J = 1$ concentration for $P = 1$ bar.

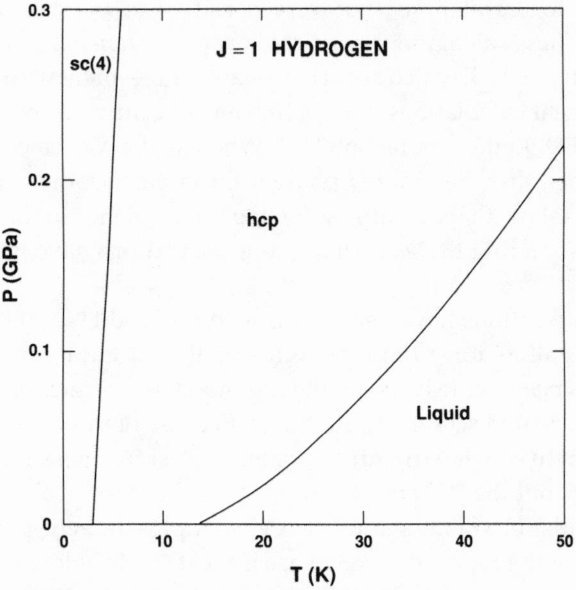

Fig. 4.3 The phase diagram of $J = 1$ H_2.

transition has recently been found in H_2 at 110 GPa and 8 K[7]. The much higher pressure of the H_2 transition follows from the smaller moment of inertia and the correspondingly higher rotational frequency of H_2.

Recent XRD work with synchrotron radiation shows that H_2 at RT up to 26 GPa is hcp, with a *c*/*a* ratio which decreases with compression[5]. This high-pressure lattice distortion may be the result of the phonon-roton interaction. Neutron diffraction has been used to determine the crystal structure of D_2 at RT up to 30.9 GPa[8]. The structure is hcp, but in this work the c/a ratio remains close to the ideal value.

A number of theoretical predictions on rotational ordering have been made[9–12]. The first significant approach was a simple mean field, which solves the quantum-mechanical problem of a perturbed rotor assuming no correlations with the orientations of the neighboring molecules. The difficult theoretical problem here is to calculate accurately the valence-electron contribution to the potential. Mean-field theory predicts a first-order phase transition from a symmetric (rotating) to a broken-symmetry (librating) phase at 37.5 GPa (H_2) and 17.5 GPa (D_2)[9]. There is a strong isotope effect due to the different moments of inertia of H_2 and D_2. A later series of calculations used a Monte Carlo variational technique to include rotational correlations[10,11]. The predicted pressures were found to be insensitive to correlations, but very sensitive to the assumed valence potential. These calculations predicted ordering transitions at ca. 120 GPa (H_2) and 40 GPa (D_2), which are in reasonable agreeement with experiment. In other recent calculations, the roton band structure has been calculated assuming EQQ interactions only[12]. Whereas the MC calculations have only assumed that the ordered phase is the cubic Pa3, the band-structure calculations show that the roton softening occurs in the part of the Brillouin zone corresponding to Pa3. All of these calculations have ignored lattice vibrations.

QMC calculations have been performed on solid H_2 at 0 K[13]. These calculations allow for proton motion as well as electron motion. A comparison has been made between the ground-state energies of the isotropic fcc(4) and oriented sc(4) Pa3 structures. Because of the very small energy difference between the structures, a precise transition pressure could not be determined, but the Pa3 structure appears to be favored above 50 GPa.

AIP calculations of the static-lattice energy plus the zero-point energy of H_2 show that the molecular Pa3 phase is most stable at low pressures, but that an ordered hcp lattice with all molecules parallel to the c-axis becomes the preferred phase at ca. 80 GPa[14].

The melting curves of H_2 and D_2 have been measured up to 373 K and 8 GPa with the piston-cylinder and DAC techniques[15–17], and the melting curve of T_2 has been measured up to 0.35 GPa in a piston-cylinder apparatus[15]. The low-pressure curves are shown in Fig. 4.4. There is a very clear isotope effect at low pressure; the heavier isotopes melt at higher temperatures. The high-pressure melting curve for H_2 is shown in Fig. 4.5.

A calculation of the H_2 and D_2 melting curves was performed with lattice dynamics and anharmonic corrections for the solid and variational-perturbation theory with quantum corrections for the liquid[18]. The pair potential was the Silvera-Goldman potential[1], a careful empirical fit to experimental data. The resulting theoretical melting curve is in good agreement with experiment for H_2. The agreement for D_2 is less good because the theory predicts the wrong sign for the isotope effect. This is a subtle problem which is not yet well understood.

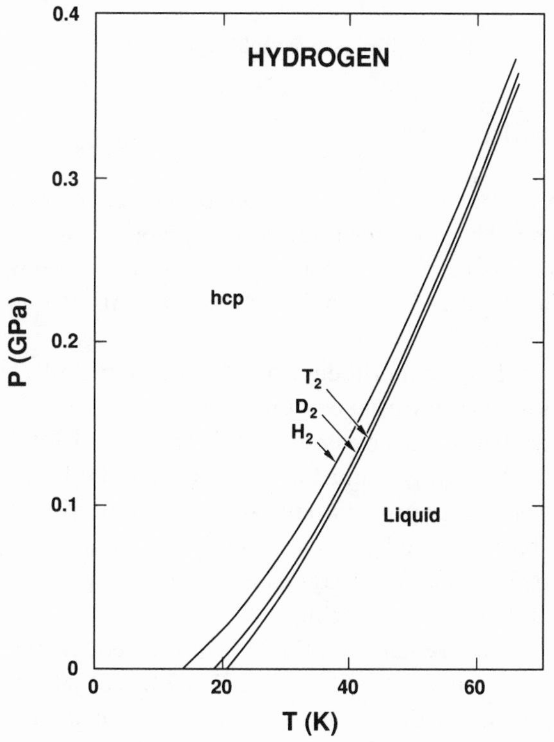

Fig. 4.4 Low-pressure melting curves for H_2, D_2, and T_2.

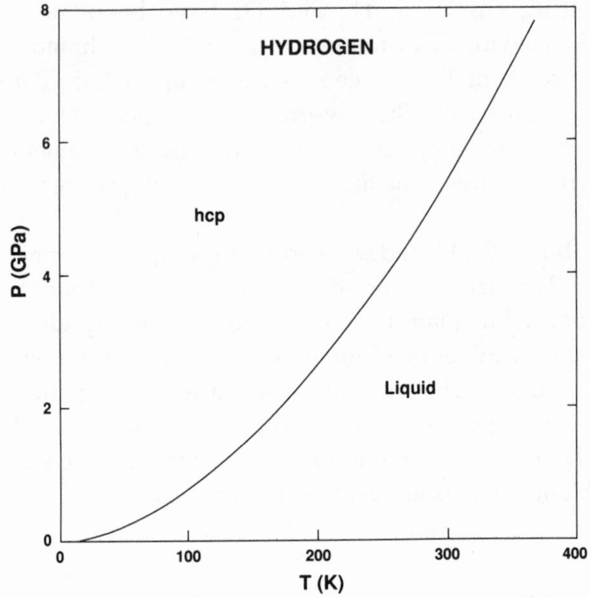

Fig. 4.5 High-pressure melting curve for H_2.

4.4 The Metallic Phase

It has been known since the 1935 calculations of Wigner and Huntington[19] that hydrogen must become metallic at very high pressure. In recent years, this fact has stimulated much experimental and theoretical work, mainly because of the expected unusual and possibly useful properties of metallic H.

There are three likely experimental approaches to metallic H: 1) isothermal compression, 2) isentropic compression, and 3) shock compression. Shock compression leads to temperatures that are too high for the formation of a degenerate electron gas and the shocked material becomes a hot H plasma. Isentropic experiments have been performed, and several experimentalists have claimed to have produced metallic H[20]. Unfortunately, only the most primitive diagnostic measurements have been made, and the claims remain unconvincing.

DAC Raman measurements on the H_2 vibron frequency show a large discontinuity in frequency at 145 GPa and 77 K[7,21,22]. A careful study of this transition has shown that it is a first-order transition to a new molecular phase, named H-A. The H-A transition boundary has a positive dP/dT and at ca. 150 K and 170 GPa, it ends in a critical point[23,24]. Reflectance measurements on H_2 to 177 GPa at RT show increasing free-electron-like behavior with compression. Analysis of the data with a Drude model suggests a continuous transition to the metallic state at 150 GPa[25]. Measurement of the molecular band gap as a function of pressure is a useful way to estimate the metallization point, but unfortunately the diamond anvils absorb photons at the same frequencies as H_2. An indirect measurement of the band gap via the index of refraction has been made to 73 GPa[26,27]. Extrapolating the index to zero frequency gives an approximate metallization pressure of 205 GPa. The experimental data to date suggest that the H-A transition is the insulator-metal transition, which proceeds by closure of an indirect gap.

In H_2 there is a second transition, apparently of the rotational-ordering type, in the H-A phase at 160–170 GPa, also with a positive dP/dT[7]. The transition equivalent to H-A in D_2 occurs at 190 GPa[28]. It is noteworthy that the isotope effect in the lower-pressure rotational-ordering transition and in the higher-pressure H-A transition have opposite signs. The structures of these new high-pressure phases have not yet been determined. A summary of the transitions in H_2 and D_2 is shown in Fig. 4.6, and the very-high-pressure phase diagram of H_2 is shown in Fig. 4.7.

It is clear from the most recent experimental work on hydrogen that the phase diagram at very high pressures is much more complex than previously imagined, and that the issue of the metallic transition will be settled by DAC technology.

Much theoretical work has been done on the molecular-to-metallic transition and on the properties of the metallic phase[20,29]. The usual procedure has been to create separate models of the insulating molecular

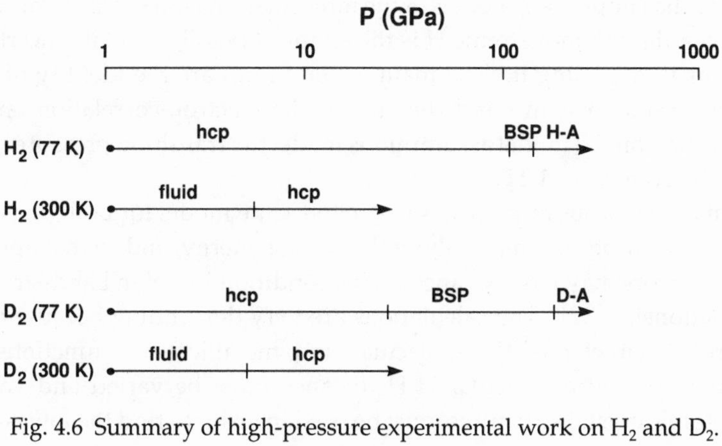

Fig. 4.6 Summary of high-pressure experimental work on H_2 and D_2.

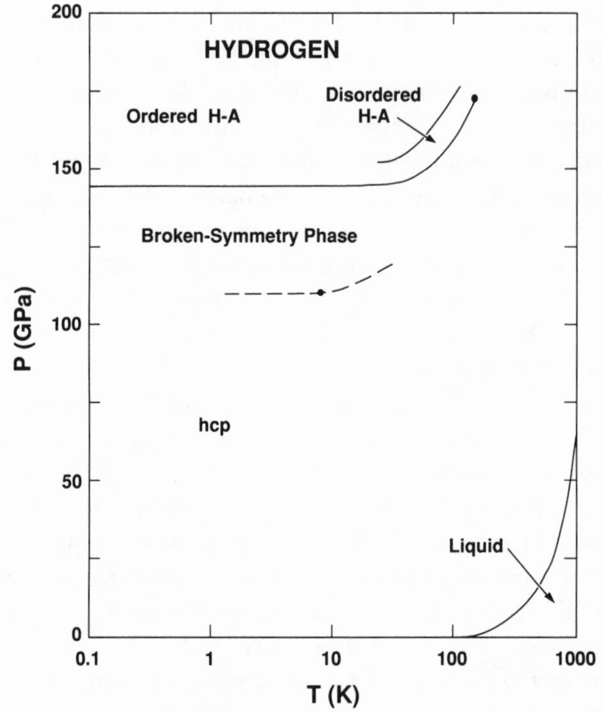

Fig. 4.7 The phase diagram of molecular H_2 at high pressures, showing the stability region of the broken-symmetry and H-A phases.

and metallic monatomic phases at 0 K and to compute the phase transition volumes and pressure by a common-tangent construction on the total energy vs. volume curves. As the quality of the experimental EOS data on solid H_2 has improved, models of the molecular phase have also improved, but even though monatomic H is the simplest possible metal, uncertainty remains in modeling it. The main uncertainties are the identity of most stable crystal structure and the size of the electron-correlation energy. Depending on the model assumptions made, the transition pressure varies roughly from 0.1 to 1 TPa.

A more rigorous approach is to use the same model for both phases, so that no assumption is made about the zero of energy, and so that approximation errors may largely cancel. This condition is met in band-structure calculations[30–32]. The calculations are very demanding because of the different character of the molecular and metallic wave functions. In addition, the intramolecular H-H distance must be varied and various monatomic crystal structures must be tried in order to find the lowest total energies of the two phases. FPLAPW calculations have been performed[31,32] in order to study the molecular-to-metallic transition. The molecular phase was assumed to be Pa3. Under compression the molecular bond increases in length as electron density became more uniform. At ca. 170 GPa, the molecular band gap closes, giving a molecular H_2 metal. At 400 $\pm$ 100 GPa, the molecular phase undergoes a first-order transition to an hcp monatomic phase. At still higher pressures near 700 GPa, the hcp phase transforms to bcc. This last transformation is driven by the dominance of the Madelung energy over the band-structure energy. Errors in these transition pressures will arise from the neglect of phonons, which are very important in H, and from the local-density approximation, which underestimates band gaps.

An argument has been made for the possibility that the monatomic phase at 0 K is a liquid rather than a solid[33]. This would depend on very different electron screening in the two phases. At ultrahigh pressures, the difference would be wiped out by the increasing electron kinetic energy, and the stable phase would then presumably be a quantum OCP solid.

The effect of nuclear motion on the structural stability of monatomic H has been studied in the NFE approximation. Static-lattice NFE calculations to third order[29] predict that isotropic lattices are unstable, but self-consistent phonon calculations reverse this conclusion and show that fcc is favored[34]. More recent AIP calculations on molecular and metallic H with harmonic zero-point motion included show transitions from cubic Pa3 to oriented-molecular hcp (m-hcp) at $P < 80$ GPa, a band-gap closure in the m-hcp phase at ca. 200 GPa, a transition to a metallic distorted simple-hexagonal phase at ca. 380 GPa, and a transition to a metallic bcc phase at ca. 850 GPa[14].

QMC calculations have also addressed the problem of the metallic transition at 0 K[13]. These calculations automatically allow for zero-point proton motion, the relaxation of the intramolecular bond distance, and the coupling of electron and proton motions. Separate calculations on diatomic (Pa3) and monatomic (fcc and bcc) lattices were performed, and a first-order transition was found at 300 $\pm$ 40 GPa. The molecular phase EOS is in moderately good agreement with experiment. It is clear that the molecular band gap is closing at high pressures, but the appearance of a molecular metal could not be confirmed. Also, the energy difference between fcc and bcc monatomic lattices was less than the numerical errors, and the most stable metallic phase could not be determined. More approximate variational QMC calculations[35] on the monatomic phase show an fcc-to-bcc transition at ultrahigh pressures.

From these contradictory theoretical results, the only conclusions that we can draw are the appearance of a molecular metal near 200 GPa and a first-order transition to a simple monatomic lattice at some higher pressure. These results emphasize the importance of choosing appropriate structures for comparison of total energies, and the importance of zero-point motion in light-atom materials like H.

For $T > 0$ the very high-pressure phase diagram is much more speculative. One interesting possibility is that a first-order diatomic-monatomic transition extends from the solid into the liquid phase and ends in a critical point[36,37]. A calculation which includes dissociation and ionization within the context of fluid perturbation theory predicts such a plasma phase transition with a critical point at 15000 K and 64.6 GPa[37]. The various suggestions for the nature of the metallization transition are summarized in Fig. 4.8.

It is also of interest to consider the melting curve of the monatomic phase at ultrahigh pressure. So far, sophisticated calculations of the melting curve have not been made. Instead, estimates based on Lindemann scaling for the quantum OCP solid have been made[38,39]. These estimates give a melting curve for H which reaches a maximum temperature of ca. 30000 K and which then bends back to a melting point at 0 K and ca. 10^7 TPa. Melting of the lattice at 0 K is due to the quantum delocalization of the protons, analogous to the Wigner transition in electrons[40]. The different quantum statistics and atomic masses of H (fermion) and D (boson) lead to different phase diagrams, as can be seen in Fig. 4.9.

4.5 Discussion

The hydrogen phase diagram is a paradigm of the interaction between theory and experiment. The hydrogen molecule and atom are the most theoretically tractable of all the elements, and this has stimulated high-precision theoretical work. Uncertainties remain because of the inherent difficulty of many-body quantum calculations. The theoretical predictions have stimulated new experimental work, and the inevitable discrepancies have led in turn to new calculations and experiments. The high quality of experimental and theoretical research on hydrogen has established a standard against which work on other elements may be usefully compared.

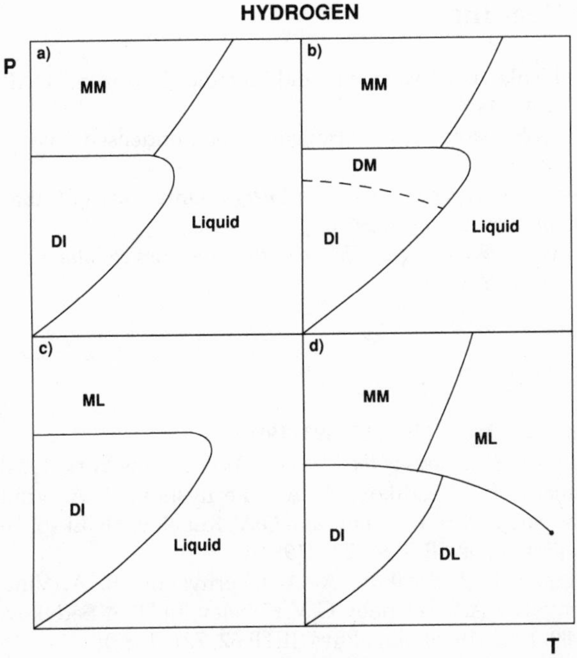

Fig. 4.8 Schematic phase diagrams proposed for the insulator-metallic phase transition in hydrogen. DI = diatomic insulator, DM = diatomic metal, MM = monatomic metal, DL = diatomic liquid, ML = monatomic liquid.

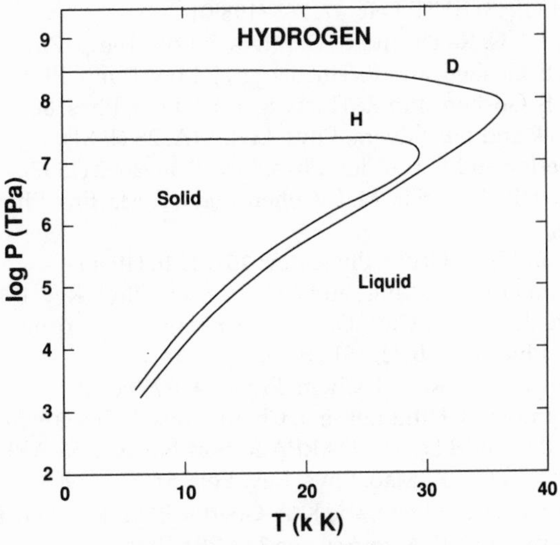

Fig. 4.9 Theoretical melting curves for metallic H and D.

For Further Reading

M. Ross and C. Shishkevish, "Molecular and Metallic Hydrogen," DARPA Report No. R-2056-ARPA, May 1977.

I. F. Silvera, "The Solid Molecular Hydrogens in the Condensed Phase,"Rev. Mod. Phys. **52**, 393 (1980).

P. C. Souers, *Hydrogen Properties for Fusion Energy* (University of California Press, Berkeley, Los Angeles, London, 1986).

J. van Kranendonk, *Solid Hydrogen: Theory of the Properties of Solid H_2, HD, and D_2* (Plenum Press, New York, 1983).

References

1. I. F. Silvera, Rev. Mod. Phys. **52**, 393 (1980).
2. J. Donohue, *The Structures of the Elements* (Wiley, New York, 1974) chap. 1.
3. S. N. Ishmaev, I. P. Sadikov, A. A. Chernyshov, B. A. Vindryaevskii, V. A. Sukhoparov, A. S. Telepnev, and G. V. Kobelev, Zh. Eksp. Teor. Fiz. **84**, 394 (1983) [Sov. Phys. JETP **57**, 228 (1983)].
4. S. N. Ishmaev, I. P. Sadikov, A. A. Chernyshov, B. A. Vindryaevskii, V. A. Sukhoparov, A. S. Telepnev, G. V. Kobelev, and R. A. Sadykov, Zh. Eksp. Teor. Fiz. **89**, 1249 (1985) [Sov. Phys. JETP **62**, 721 (1985)].
5. H. K. Mao, A. P. Jephcoat, R. J. Hemley, L. W. Finger, C. S. Zha, R. M. Hazen, and D. E. Cox, Science **239**, 1131 (1988).
6. I. F. Silvera and R. J. Wijngaarden, Phys. Rev. Lett. **47**, 39 (1981).
7. H. E. Lorenzana, I. F. Silvera, and K. A. Goettel, Phys. Rev. Lett. **64**, 1939 (1990).
8. V. P. Glazkov, S. P. Besedin, I. N. Goncharenko, A. V. Irodova, I. N. Makarenko, V. A. Somenkov, S. M. Stishov, and S. Sh. Shil´shtein, Pis´ma Zh. Eksp. Teor. Fiz. **47**, 661 (1988) [JETP Lett. **47**, 763 (1988)].
9. W. England, J. C. Raich, and R. D. Etters, J. Low Temp. Phys. **22**, 213 (1976).
10. I. Aviram, S. Goshen, and R. Thieberger, J. Low Temp. Phys. **52**, 397 (1983).
11. I. Aviram, S. Goshen, and R. Thieberger, J. Chem. Phys. **80**, 5337 (1984).
12. A. Lagendijk and I. F. Silvera, Phys. Lett. **84A**, 28 (1981).
13. D. M. Ceperley and B. J. Alder, Phys. Rev. B **36**, 2092 (1987).
14. T. W. Barbee III, A. Garcia, M. L. Cohen, and J. L. Martins, Phys. Rev. Lett. **62**, 1150 (1989).
15. R. L. Mills and E. R. Grilly, Phys. Rev. **101**, 1246 (1956).
16. D. H. Liebenberg, R. L. Mills, and J. C. Bronson, Phys. Rev. B **18**, 4526 (1978).
17. V. Diatschenko, C. W. Chu, D. H. Liebenberg, D. A. Young, M. Ross, and R. L. Mills, Phys. Rev. B **32**, 381 (1985).
18. D. A. Young and M. Ross, J. Chem. Phys. **74**, 6950 (1981).
19. E. Wigner and H. B. Huntington, J. Chem. Phys. **3**, 764 (1935).
20. M. Ross and C. Shishkevish, DARPA Report No. R-2056-ARPA (1977).
21. R. J. Hemley and H. K. Mao, Phys. Rev. Lett. **61**, 857 (1988).
22. H. E. Lorenzana, I. F. Silvera, and K. A. Goettel, Phys. Rev. Lett. **63**, 2080 (1989).
23. R. J. Hemley and H. K. Mao, Science **249**, 391 (1990).
24. H. E. Lorenzana, I. F. Silvera, and K. A. Goettel, Phys. Rev. Lett. **65**, 1901 (1990).

25. H. K. Mao, R. J. Hemley, and M. Hanfland, Phys. Rev. Lett. **65**, 484 (1990).
26. J. H. Eggert, K. A. Goettel, and I. F. Silvera, Europhys. Lett. **11**, 775 (1990).
27. J. H. Eggert, K. A. Goettel, and I. F. Silvera, Europhys. Lett. **12**, 381 (1990).
28. R. J. Hemley and H. K. Mao, Phys. Rev. Lett. **63**, 1393 (1989).
29. A. K. McMahan, in *High-Pressure and Low-Temperature Physics*, C. W. Chu and J. A. Woollam, eds. (Plenum, New York, 1978) p. 21.
30. D. E. Ramaker, L. Kumar, and F. E. Harris, Phys. Rev. Lett. **34**, 812 (1975).
31. B. I. Min, H. J. F. Jansen, and A. J. Freeman, Phys. Rev. B **30**, 5076 (1984).
32. B. I. Min, H. J. F. Jansen, and A. J. Freeman, Phys. Rev. B **33**, 6383 (1986).
33. A. H. MacDonald and C. P. Burgess, Phys. Rev. B **26**, 2849 (1982).
34. D. M. Straus and N. W. Ashcroft, Phys. Rev. Lett. **38**, 415 (1977).
35. D. Ceperley, G. V. Chester, and M. H. Kalos, Phys. Rev. B **16**, 3081 (1977).
36. W. Ebeling and W. Richert, Phys. Lett. **108A**, 80 (1985).
37. D. Sauman and G. Chabrier, Phys. Rev. Lett. **62**, 2397 (1989).
38. R. Mochkovitch and J. P. Hansen, Phys. Lett. **73A**, 35 (1979).
39. H. Kawamura and K. Ebina, Phys. Lett. **103A**, 273 (1984).
40. D. M. Ceperley and B. J. Alder, Phys. Rev. Lett. **45**, 566 (1980).

CHAPTER 5
The Group I Elements (The Alkali Metals)

5.1 Introduction

The alkali metals are of technological interest primarily as liquid coolants for nuclear reactors. Much of the experimental research on these metals has been motivated by this interest.

Because of the high compressibilities and low melting points of the alkalis, their phase diagrams have been determined over a significant range of compressions. These studies reveal complex diagrams with many crystal structures and surprising anomalies in the melting curves.

This experimental work has attracted the attention of theorists, and because the alkalis at low pressure are NFE metals, a very large body of theoretical work using the NFE pseudopotential approximation has been done on the thermodynamic properties, including the phase diagrams, of these metals. At higher pressures in the heavier alkalis, the valence electrons shift from largely *s* character to *d* character. This *s*-*d* electron transfer gives rise to the unusual melting curves and solid-solid phase transitions. Band-structure calculations have given us a good understanding of this phenomenon, but more work is needed to understand the precise relationship between the *s*-*d* transfer and the phase diagrams.

5.2 Lithium

At RTP Li is bcc. As Li is cooled below about 75 K, it transforms to a close-packed structure, originally considered to be hcp[1]. Cold working of the solid below 75 K produces fcc[1]. The bcc to close-packed (cp) transition is martensitic and shows strong hysteresis in both temperature and structure. More recent neutron-diffraction studies show that the new structure below 75 K is the close-packed samarium-type hex(9) or rh(3) lattice[2–6]. This structure has the stacking sequence ABABCBCAC or chhchhchh, and was originally thought to be restricted to the lanthanides (see chapter 14). The hex(9) phase begins to nucleate in the bcc matrix below 75 K, and it increases in concentration as the temperature is decreased. This phase has a significant concentration of stacking faults[4,5].

Resistance measurements at high pressure and low temperatures show a discontinuity at 7 K and 26 GPa[7]. Analysis of the resistance curves suggests that this is a crystallographic phase change, but the phases have not been identified. No further transition is seen up to 41 GPa[7].

The bcc-cp phase boundary has been determined by an acoustic-emission technique to ca. 3 GPa, and it has a positive dP/dT[6,8]. On warming at 0.65 GPa, the hex(9) phase transforms to fcc over a narrow temperature range[9]. At RT and 6.9 GPa, Li transforms from bcc to fcc[10]. This phase transition has not been followed as a function of temperature, but the positive slope of the bcc-cp boundary at low temperatures strongly suggests that the RT transition is a continuation of the low-temperature transition, and that the transition should reach a triple point on the melting curve above 10 GPa. It is clear from the conflicting evidence that the equilibrium phase boundaries have not yet been determined at low temperatures, but there may be a bcc-fcc-hex(9) triple point at low temperature and pressure. Compression of Li at RT to 10 GPa shows no further transitions[10].

Because Li is one of the simplest metals, it has received much theoretical attention. QMC calculations have been performed, and they are consistent with band-structure results[11]. Numerous static-lattice total-energy calculations have been made. LMTO[12,13], ASW[14], LCGTO[15], KKR[16], and AIP[17] calculations all predict that a close-packed phase is favored over bcc at $P = 0$. The calculations differ in the location of transitions to other phases, but the most extensive calculation[13] finds the sequence hcp→fcc→hcp→bcc, with the last transition occurring at 2.6 TPa. Hartree-Fock calculations of clusters of Li atoms, including the fcc, hcp, bcc, and hex(9) structures, clearly indicate that the hex(9) has the lowest energy[18].

Band-structure calculations for very high compressions show a band reordering in which the $2s$ valence electron takes on increasing $2p$ character with increasing pressure[13,14]. The result is a flattening or "softening" of the pressure-density isotherm in the 1 TPa region. In the heavier alkali metals the corresponding shift of valence electrons from s to d character has a strong effect on the phase diagram.

NFE pseudopotential theory is not an accurate description of Li, because the p-electron component of the valence band is not excluded from the $1s^2$

atomic core, and therefore it sees a strong potential which cannot be treated as a perturbation. Nevertheless, theoretical calculations using NFE pseudopotential theory do indeed predict a bcc-to-close-packed phase transition at low temperature, in agreement with experiment. Most of these calculations show a negative slope to the phase boundary[6,19,20], but molecular dynamics calculations using a NFE pair potential predict a positive slope, as is seen experimentally[21].

The melting curve of Li has been measured to 8 GPa[22,23]. At high pressure it shows a nearly infinite dP/dT, which might be related to the bcc-fcc-liquid triple point anticipated at still higher pressures. Calculations of the melting curve using pseudopotential theory are in adequate agreement with the experimental curve[20,24]. Quantum corrections for the liquid are necessary for accurate results. The phase diagram of Li is shown in Fig. 5.1.

5.3 Sodium

At RTP Na is bcc. When cooled below 36 K, Na transforms, like Li, by a martensitic transition to a phase originally thought to be hcp[1]. Neutron-diffraction evidence suggests that, on the contrary, the same faulted hex(9)

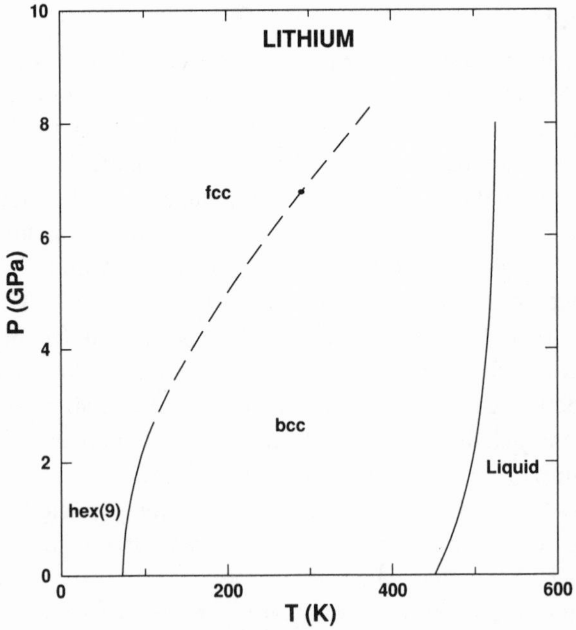

Fig. 5.1 The phase diagram of lithium.

lattice found in Li also appears in Na[5,6]. The bcc-cp phase boundary has been measured[6,8,25], and is found to have a negative slope, with a 0 K intercept at between 0.1 and 0.2 GPa. These measurements are more difficult and less certain than for lithium. Compression of Na to 27 GPa at RT shows no further phase transitions[26].

LMTO, AIP, and GPT static-lattice total-energy calculations have been performed on Na for the bcc, fcc, and hcp structures. LMTO[6,12] incorrectly predicts bcc stability at $T = 0$ and $P = 0$, while AIP[17] and GPT[27] predict hcp stability. The AIP and GPT calculations both predict an hcp-bcc transition at ca. 1 GPa, which is in good agreement with experiment, considering the very small energy differences between phases. At much higher pressures, GPT predicts a transition from bcc back to hcp near 100 GPa due to the approach of the 3*d* band to the 3*s* valence band[27]. So far, there is no experimental confirmation of this prediction. NFE pseudopotential theory with lattice dynamics has been successful in predicting the low-temperature cp stability and it gives a semiquantitative prediction of the phase boundary[19]. Na is one of the best examples of a NFE metal, and it is expected that the pseudopotential theory will do well in predicting thermodynamic properties. However, it is surprising and encouraging that the theory can predict a bcc-cp phase transition, which is driven by exceedingly small energy differences, about 10^{-4} of the valence-electron binding energy. This transition results from a subtle balancing of Madelung, band-structure, and phonon energies[6].

The melting curve of Na has been measured to 12 GPa in a DAC by visual observation of the change in shape of the sample upon melting[28]. A number of theorists have made calculations of the melting curve using accurate models of the solid and fluid phases[20,24,29,30]. In general, these calculations show semiquantitative agreement with experiment. The pseudopotential calculations for melting support the NFE model and in addition demonstrate the high quality of the statistical-mechanical models used for solid and liquid Na. Comparison of experiment and theory for Na is shown in Fig. 5.2. The phase diagram of Na is shown in Fig. 5.3.

5.4 Potassium

At RTP, K is bcc. At low temperatures there is evidence for a charge-density-wave distortion of the bcc lattice[31], but no clear phase transition has been observed[5]. Recent DAC XRD[32] and reflectivity[33] measurements have located rather precisely a series of phase transitions along the RT isotherm. According to XRD measurements, at 11.6 GPa there is a bcc-fcc transition

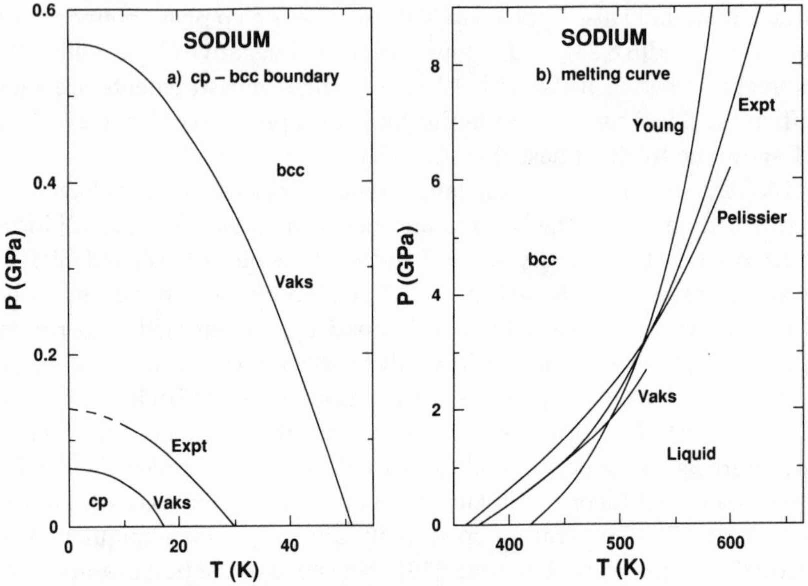

Fig. 5.2 Comparison of theory and experiment for sodium.

and at ca. 20.0 GPa there is another transition to a new, unknown structure, K III. From discontinuities in optical reflectivity, these pressures are found to be 11.4 and 18.8 GPa. No further changes are found up to 38.0 GPa[33].

At $P = 0$, LMTO[12,34] and AIP[17] calculations both predict bcc stability for K, as is observed. At higher pressures, LMTO predicts the sequence bcc→hcp→fcc. The hcp phase has not been observed in K, but the predicted hcp→fcc transition pressures[12,34] of 10.7 and 11.0 GPa are in good agreement with the observed bcc→fcc transition at RT.

It is clear from band-structure calculations that the transitions in K are due to the *s-d* electron transfer[35,36]; specifically they are the result of the approach and overlap of the half-filled 4*s* valence band and the empty 3*d* band. The higher-angular-momentum *d* band has lower kinetic energy than the *s* band, and the *d* band will therefore increase in energy more slowly with compression than the *s* band. The result is that at sufficiently high compression, the two bands will overlap and the valence electrons will increasingly take on *d* character. According to ASW calculations, this occurs near 50 GPa, producing a strong flattening of the 0 K isotherm[36]. A number of attempts have been made to understand the systematics of the bcc-to-fcc transitions in the heavier alkali metals, and the most successful relate this transition to the progress of the *s-d* transfer[12,37]. This transition correlates with the fraction of *d* character in the valence band, or

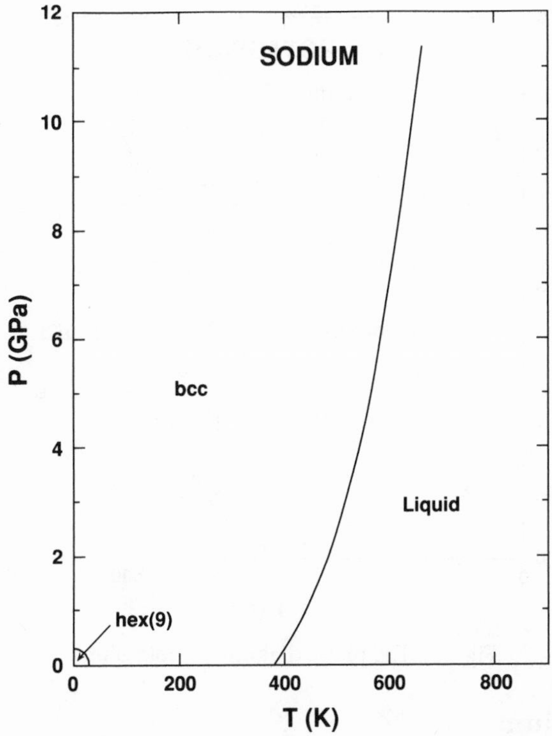

Fig. 5.3 The phase diagram of sodium.

equivalently to a constant ratio of atomic to ionic radius[32], $R_A/R_I \cong 1.5$. The higher-pressure K III phase, like the corresponding phases in Rb and Cs, is probably a more open structure which is stabilized by the more localized, covalent-like character of the *d* component of the valence band.

LMTO band-structure calculations on K predict the end of the *s*-*d* transfer by 60 GPa, followed by a sequence of phase transitions hcp→hex(4) (dhcp)→hex(9)→bcc, occurring at pressures of 280, 810, and 2100 GPa[38]. This structural sequence appears to be dominated by ionic-core repulsion in the "transition-metal" K.

The melting curve of K has been measured in the DAC to 14.5 GPa[28]. The bcc-fcc-liquid triple point is found at 11.0 GPa, showing that the bcc-fcc phase boundary has a nearly zero dP/dT. Also near the triple point, the melting curve has a nearly infinite dP/dT, indicating a rapid densification of the liquid which is interpreted to be the result of the continuous *s*-*d* transfer in the liquid phase. Theoretical calculations of melting based on NFE pseudopotential theory are in good agreement with experiment[20,24,39]. The phase diagram of K is shown in Fig. 5.4.

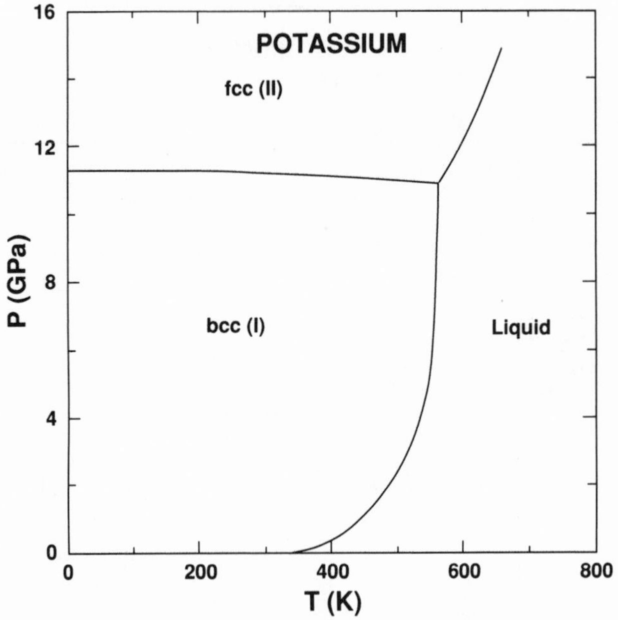

Fig. 5.4 The phase diagram of potassium.

5.5 Rubidium

At RTP, Rb is bcc. There is some evidence for a phase transition at low temperature, but this is not yet convincingly demonstrated[40]. DAC XRD measurements along the RT isotherm reveal a very complex series of phase transitions[32]. At 7 GPa, the bcc (I) phase transforms to fcc (II). Two new, unidentified phases, Rb III and Rb IV, appear at 14 GPa and 17 GPa. At ca. 20 GPa there is another transition to Rb V, which has been indexed to the tetragonal ct(4) Cs IV lattice. From near-infrared reflectivity discontinuities, these transition pressures are found to be 7.0, 12.8, 16.0, and 19.0 GPa[41]. No further transitions are observed up to 35 GPa[32].

The bcc-fcc phase boundary has been measured from RT to the melting curve[42]. The slope is positive and surprisingly large in comparison to the same transition in K and Cs. Very approximate measurement of the II-III boundary suggests that $dP/dT \cong 0$ in this case.

As with K, the complex crystal structures in Rb are thought to be results of the *s*-*d* transfer. The bcc-fcc transition occurs at $R_A/R_I \cong 1.5$, as in K and Cs[32]. AIP calculations on Rb predict a bcc-fcc transition at 5.2 GPa[43], which is in good agreement with the extrapolation of the experimental data to 0 K. This theoretical study ascribes the transition to the contribution of the nonlocal *d*-electron pseudopotential, which is consistent with the transition being related to the *s*-*d* transfer.

LMTO calculations for Rb predict the end of the *s*-*d* transfer at 53 GPa, followed by the transitions hcp→hex(9)→bcc, at 310 and 1000 GPa, respectively[38].

The melting curve of Rb has been measured to 14 GPa[22,42]. Unusual structure is seen in the vicinity of the bcc-fcc-liquid and fcc-Rb III-liquid triple points. This follows from the *s*-*d* transfer occurring continuously in the liquid phase, which can become more dense than the solid, and thus give $dP/dT < 0$. Although this phenomenon can be predicted qualitatively from pair potentials that exhibit "softening" (see chapter 3), there have not yet been any quantitative predictions of the Rb melting curve in the *s*-*d* region. NFE Pseudopotential calculations give adequate agreement with experiment only in the low-pressure region[24,39]. The phase diagram of Rb is shown in Fig. 5.5.

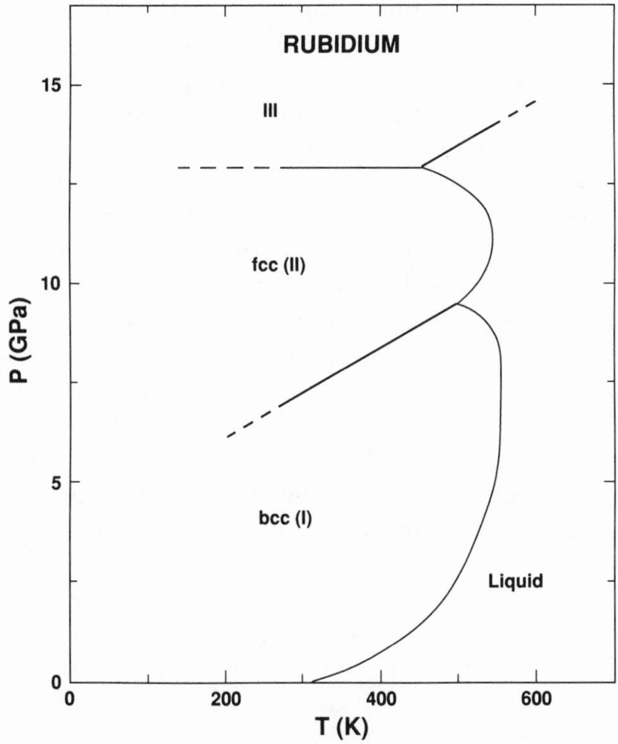

Fig. 5.5 The phase diagram of rubidium.

5.6 Cesium

At RTP, Cs is bcc (I). There is no evidence for any anomalous behavior at low temperatures[1]. At RT and 2.25 GPa, Cs transforms to Cs II, which is fcc[44,45]. At 4.2 GPa, there is a transition to a second fcc phase, III, with a 9% volume decrease[44]. At 4.3 GPa, there is a transition to the ct(4) tetragonal phase IV, in which each atom has 8 nearest neighbors[46]. At ca. 10.0 GPa, there is another transition to Cs V, which has an uncertain structure, either simple tetragonal or simple orthorhombic, with sixfold coordination[47]. At ca. 72 GPa, Cs V transforms to Cs VI, which is hcp or dhcp[48]. No further transitions are observed up to 92 GPa[48].

The Cs I-II phase boundary has been measured from 77 K to a triple point on the melting curve near 460 K[45]. The II-III and III-IV transitions have been measured from the melting curve to ca. 270 K, where they become sluggish and phase II can no longer be detected[49,50]. There may be a II-III-IV triple point near 270 K. All of these transitions show rather small values of dP/dT.

Because these unusual phase transitions occur at relatively low pressures, they have been intensively studied, both experimentally and theoretically. Our current understanding of the Cs phase diagram is good, but there are many detailed features which still need study. The bcc-fcc transition correlates with those in K and Rb, and has a ratio $R_A/R_I \cong 1.5$[32]. As with K and Rb, it appears that the transition occurs when the valence electrons acquire a fixed fraction of *d* character. This has been demonstrated for Rb and Cs by measurement of interband absorption at high pressure[51].

The 4.2 GPa II-III transition has excited the most interest, because it is isostructural (fcc to fcc), and must therefore indicate some kind of electronic rearrangement in the Cs atom[52]. Band-structure calculations have revealed a rather complete picture of the *s-d* transition in this range and convincingly explain the isostructural transition[53,54]. In Cs at low pressure, the 5*d* band initially lies above the 6*s* band, and compression causes the two to converge, as shown in Fig. 5.6. The approach of the two bands causes hybridization and increasing *d* character in the valence band[55]. This accounts for the high compressibility of Cs as well as the bcc-fcc transition. At 4.2 GPa, a branch of the 5*d* band which cannot hybridize with the 6*s* band because of symmetry suddenly crosses the Fermi level and abruptly softens the isotherm. By itself, this is not enough to drive a first-order phase transition, but the softening also gives rise to a minimum in the Grüneisen parameter and hence a minimum in the thermal pressure. Measurements of the Grüneisen parameter for Cs do indeed show an anomalous decrease near the transition[56], which supports this interpretation. The thermal pressure minimum creates a van der Waals loop in the isotherm, which indicates a first-order isostructural phase transition. The LMTO-predicted transition pressure at 4.1 GPa is very close to the experimental value, as shown in Fig. 5.7. As in the bcc-fcc transition, the ratio R_A/R_I appears to be constant[32], about 1.4 for the II-III transitions in K, Rb, and Cs, again suggesting a fixed point in the *s-d* transfer process.

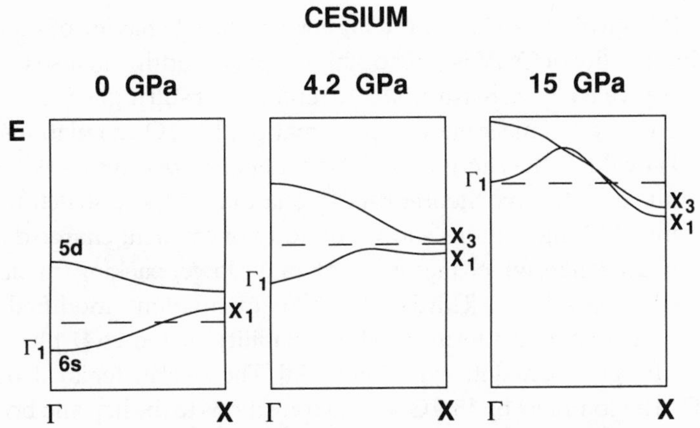

Fig. 5.6 The band structure of cesium as a function of pressure.

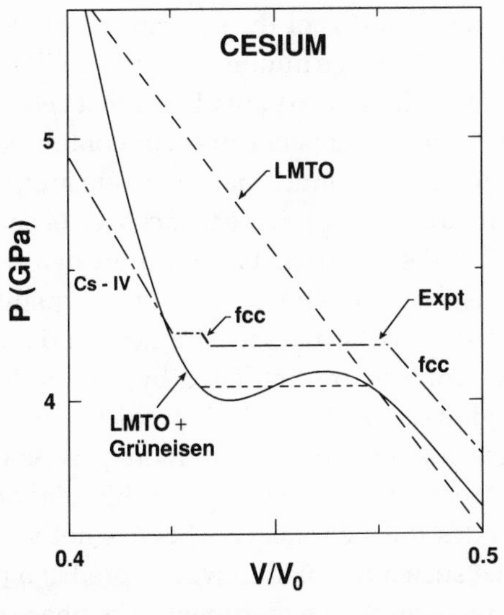

Fig. 5.7 The experimental and theoretical isotherms of cesium in the isostructural-transition region.

There remains an interesting theoretical problem regarding the Cs II-Cs III phase line. The Cs III phase could disappear in a triple point or in a critical point. The latter possibility is suggested by the requirement that thermal effects drive the II-III transition. At low enough temperatures, the transition can no longer be driven, and it must either end in a critical point or be terminated in a triple point by intersection with the III-IV phase boundary. Further theoretical and experimental work on this question is needed.

LMTO calculations[38] on the higher-pressure behavior of Cs indicate that the stability of Cs IV is due to a higher band width, but also that this is rapidly canceled by increasing 5*p* core repulsion at still higher compression. Because of the atomic-sphere approximation, LMTO calculations cannot properly evaluate the energies of the more open structures Cs IV and Cs V, so that the quantitative theoretical explanation for these structures is still uncertain. AIP calculations show a buildup of covalent charge density in this pressure range, which may account for the lower packing fraction of the IV and V phases[57]. Third-order NFE calculations modified with a repulsive core-overlap term do show stability of the ct(4) phase above 4.2 GPa, in agreement with experiment[58]. The *s-d* transfer is calculated by LMTO to be complete by 15.0 GPa, and transitions to the hcp and bcc phases are then predicted[38]. The appearance of hcp (or dhcp) at 72 GPa confirms this[48]. The hcp-bcc pressure is predicted to be 220 GPa.

The melting curve of Cs has been accurately measured[42,49,59], and it dramatically shows the progress of the *s-d* transfer. There are two small maxima near 2.2 GPa, and a large minimum near 4.2 GPa. Above 4.3 GPa, dP/dT is positive. The qualitative explanation is that because of the atomic disorder in the liquid, the *s-d* transfer can occur continuously over a wide pressure range, while the solid must make discrete jumps in volume and structure. Major structural changes have in fact been observed in liquid Cs up to 4.3 GPa[60]. Thus the liquid can become more dense than the solid in the region below the isostructural transition. A more quantitative explanation takes note of the negative Grüneisen parameter and the implications of this for the melting temperature predicted by the Lindemann Law[61]. There is indeed a region of negative dP/dT predicted as a result of the behavior of the Grüneisen parameter. A still more precise theoretical treatment requires calculation of the free energy of the solid and liquid with a model that incorporates the *s-d* transfer. This has not yet been done, but there is no doubt that such a calculation would predict a melting curve in semiquantitative agreement with experiment. The phase diagram of Cs is shown in Fig. 5.8.

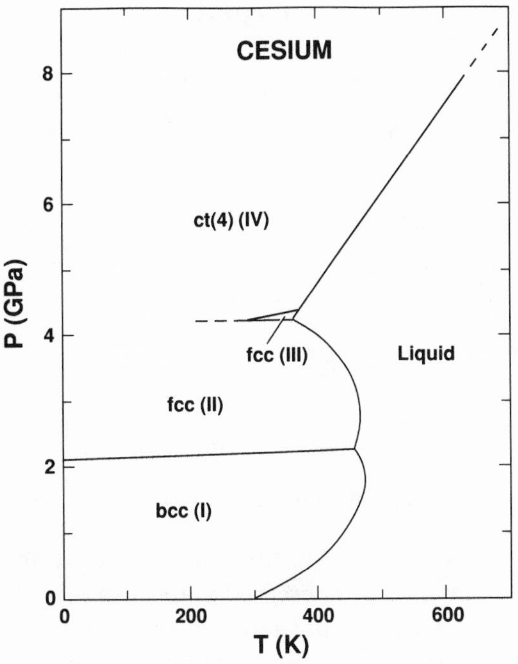

Fig. 5.8 The phase diagram of cesium.

5.7 Francium

Fr is a highly radioactive element. The longest-lived isotope has a half-life of only 22 min, so it is unlikely that experimental work on the purified element will ever be done. However, high-speed chemical studies on aqueous Fr^+ ions suggest that Fr is a normal alkali metal[62].

5.8 Discussion

Because of their NFE character at low pressures and the *s-d* transfer at higher pressures, the alkali metals are ideal subjects for theoretical study. Their melting curves and solid-solid phase transitions are within reach of current experimental techniques, so theory can be confronted with accurate measurements. Since *s-d* transfer and related band-crossing phenomena are universal among the elements at ultrahigh pressures[12,63,64], the alkalis are useful as experimentally accessible examples of this phenomenon. The high-pressure structures of the alkali metals at RT are shown in Fig. 5.9.

The elements K, Rb, and Cs show a kind of "corresponding-states" pattern in the location of the bcc-fcc transition as indicated by the

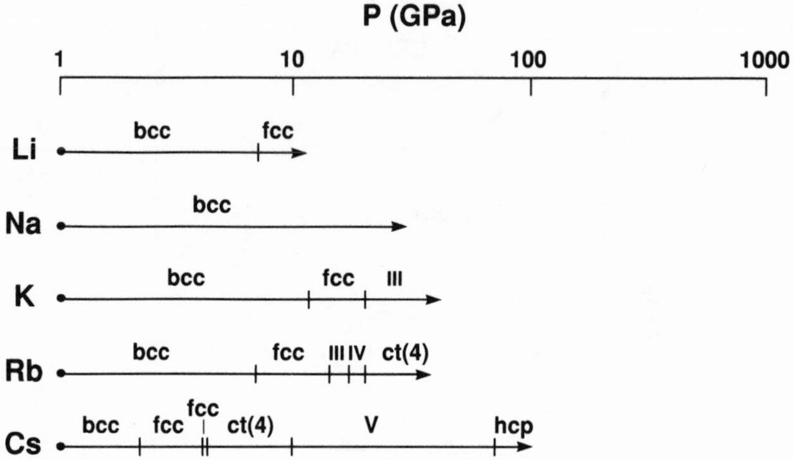

Fig. 5.9 RT transition pressures for the alkali metals.

dimensionless ratio R_A/R_I[32]. The trend of decreasing transition pressure with increasing atomic weight observed in the alkalis is also widespread among the elements. This is simply understood in atomic-physics terms. In heavy members of a group of elements, the valence electrons see the same charge as the lighter members, but they are farther from the nucleus because the ionic core is larger[65]. This is illustrated in Fig. 5.10. Hence the electron-binding energies and level spacings are smaller. The same is true of electron bands in the solid, and so for heavier elements it takes less pressure to effect a band crossing or other change which drives the phase transition.

That the heavier alkalis do not show the same sequence of phases at all pressures indicates that the band structures of these elements, although similar, are not precisely analogous, so that "corresponding states" does not rigorously hold true.

For Further Reading

R. W. Ohse, ed., *Handbook of Thermodynamic and Transport Properties of Alkali Metals* (Blackwell, Oxford, 1985).

References

1. J. Donohue, *The Structures of the Elements* (Wiley, New York, 1974) chap. 3.
2. A. W. Overhauser, Phys. Rev. Lett. **53**, 64 (1984).
3. H. G. Smith, Phys. Rev. Lett. **58**, 1228 (1987).

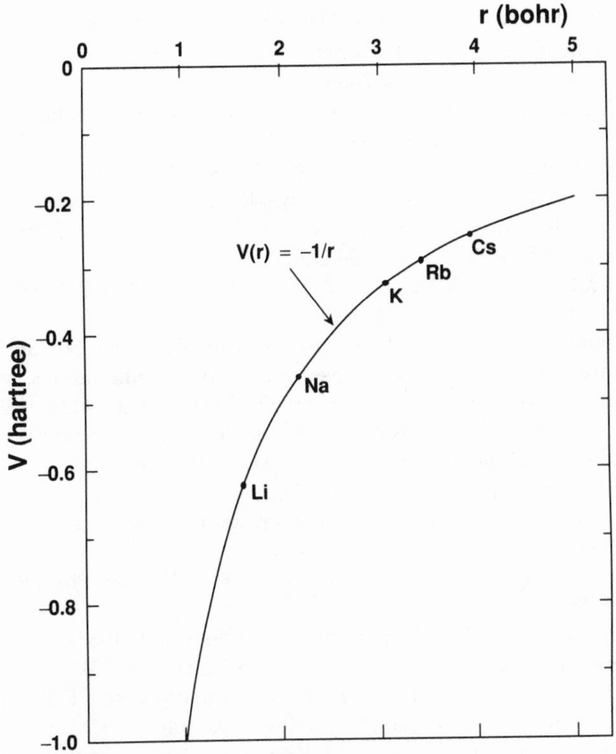

Fig. 5.10 Typical pseudopotential core parameters r_c superimposed on the $Z_i = 1$ Coulomb potential for alkali metals.

4. R. Berliner and S. A. Werner, Phys. Rev. B **34**, 3586 (1986).
5. R. Berliner, O. Fajen, H. G. Smith, and R. L. Hitterman, Phys. Rev. B **40**, 12086 (1989).
6. V. G. Vaks, M. I. Katsnelson, V. G. Koreshkov, A. I. Likhtenstein, O. E. Parfenov, V. F. Skok, V. A. Sukhoparov, A. V. Trefilov, and A. A. Chernyshov, J. Phys: Condens. Matter **1**, 5319 (1989).
7. T. H. Lin and K. J. Dunn, Phys. Rev. B **33**, 807 (1986).
8. A. A. Chernyshov, V. A. Sukhoparov, and R. A. Sadykov, JETP Lett. **37**, 405 (1983) [Pis'ma Zh. Eksp. Teor. Fiz. **37**, 345 (1983)].
9. H. G. Smith, R. Berliner, and J. D. Jorgensen, Physica B **156&157**, 53 (1989).
10. B. Olinger and J. W. Shaner, Science **219**, 1071 (1983).
11. G. Sugiyama, G. Zerah, and B. J. Alder, in *Strongly Coupled Plasma Physics*, F. J. Rogers and H. E. DeWitt, eds. (Plenum, New York, 1987) p. 229.
12. H. L. Skriver, Phys. Rev. B **31**, 1909 (1985).
13. J. C. Boettger and R. C. Albers, Phys. Rev. B **39**, 3010 (1989).
14. W. G. Zittel, J. Meyer-ter-Vehn, J. C. Boettger, and S. B. Trickey, J. Phys. F **15**, L247 (1985).
15. J. C. Boettger and S. B. Trickey, Phys. Rev. B **32**, 3391 (1985).

16. S. V. Chernov, High Temp. 26, 191 (1988) [Teplofiz. Vys. Temp. **26**, 264 (1988)].
17. M. M. Dacorogna and M. L. Cohen, Phys. Rev. B **34**, 4996 (1986).
18. A. D. Zdetsis, Phys. Rev. **34**, 7666 (1986).
19. V. G. Vaks, S. P. Kravchuk, and A. V. Trefilov, Sov. Phys. Solid State **19**, 1983 (1977) [Fiz. Tverd. Tela **19**, 3396 (1977)].
20. D. A. Young and M. Ross, Phys. Rev. B **29**, 682 (1984).
21. R. G. Munro and R. D. Mountain, Phys. Rev. B **28**, 2261 (1983).
22. H. D. Luedemann and G. C. Kennedy, J. Geophys. Res. **73**, 2795 (1968).
23. R. Boehler, Phys. Rev. B **27**, 6754 (1983).
24. A. M. Bratkovskii, V. G. Vaks, and A. V. Trefilov, Sov. Phys. JETP **59**, 1245 (1984) [Zh. Eksp. Teor. Fiz. **86**, 2141 (1984)].
25. H. G. Smith, R. Berliner, and J. Trivisonno, Bull. Am. Phys. Soc. **35**, 575 (1990).
26. I. V. Aleksandrov, V. N. Kachinskii, I. N. Makarenko, and S. M. Stishov, Pis´ma Zh. Eksp. Teor. Fiz. **36**, 336 (1982) [JETP Lett. **36**, 411 (1982)].
27. A. K. McMahan and J. A. Moriarty, Phys. Rev. B **27**, 3235 (1983).
28. C.-S. Zha and R. Boehler, Phys. Rev. B **31**, 3199 (1985).
29. J. L. Pelissier, Physica **121A**, 217 (1983).
30. B. L. Holian, G. K. Straub, R. E. Swanson, and D. C. Wallace, Phys. Rev. B **27**, 2873 (1983).
31. T. M. Giebultowicz, A. W. Overhauser, and S. A. Werner, Phys. Rev. Lett. **56**, 1485 (1986).
32. H. Olijnyk and W. B. Holzapfel, Phys. Lett. **99A**, 381 (1983).
33. K. Takemura and K. Syassen, Phys. Rev. B **28**, 1193 (1983).
34. M. Alouani, N. E.Christensen, and K. Syassen, Phys. Rev. B **39**, 8096 (1989).
35. M. Ross and A. K. McMahan, Phys. Rev. B **26**, 4088 (1982).
36. W. Zittel, J. Meyer-ter-Vehn, and J. Köbler, Solid State Comm. **62**, 97 (1987).
37. T. M. Yeremenko and Ye. V. Zarochentsev, Phys. Met. Metall. **52**, 1 (1981) [Fiz. Metal. Metalloved. **52**, 7 (1981)].
38. A. K. McMahan, Phys. Rev. B **29**, 5982 (1984).
39. J. L. Pelissier, Physica **126A**, 474 (1984).
40. I. M. Templeton, J. Phys. F **12**, L121 (1982).
41. K. Takemura and K. Syassen, Solid State Comm. **44**, 1161 (1982).
42. R. Boehler and C.-S. Zha, Physica **139&140B**, 233 (1986).
43. W. Maysenhölder, S. G. Louie, and M. L. Cohen, Phys. Rev. B **31**, 1817 (1985).
44. H. T. Hall, L. Merrill, and J. D. Barnett, Science **146**, 1297 (1964).
45. M. S. Anderson, E. J. Gutman, J. R. Packard, and C. A. Swenson, J. Phys. Chem. Solids **30**, 1587 (1969).
46. K. Takemura, S. Minomura, and O. Shimomura, Phys. Rev. Lett. **49**, 1772 (1982).
47. K. Takemura and K. Syassen, Phys. Rev. B **32**, 2213 (1985).
48. K. Takemura, O. Shimomura, and H. Fujihisa, preprint.
49. A. Jayaraman, R. C. Newton, and J. M. McDonough, **159**, 527 (1967).
50. D. B. McWhan and A. L. Stevens, Solid State Comm. **7**, 301 (1969).
51. H. Tups, K. Takemura, and K. Syassen, Phys. Rev. Lett. **49**, 1776 (1982).
52. R. Sternheimer, Phys. Rev. **78**, 235 (1950).
53. A. K. McMahan, Phys. Rev. B **17**, 1521 (1978).
54. D. Glötzel and A. K. McMahan, Phys. Rev. B **20**, 3210 (1979).
55. A. K. McMahan, Int. J. Quantum Chem. Symp. **20**, 393 (1986).

56. R. Boehler and M. Ross, Phys. Rev. B **29**, 3673 (1984).
57. S. G. Louie and M. L. Cohen, Phys. Rev. B **10**, 3237 (1974).
58. T. Soma and H. Kagaya, J. Phys. F **17**, L1 (1987).
59. G. C. Kennedy, A. Jayaraman, and R. C. Newton, Phys. Rev. **126**, 1363 (1962).
60. K. Tsuji, K. Yaoita, M. Imai, T. Mitamura, T. Kikegawa, O. Shimomura, and H. Endo, J. Non-Cryst. Solids **117/118**, 72 (1990).
61. S. N. Vaidya, High Temp.-High Press. **11**, 335 (1979).
62. S. H. Eberle, H. W. Kirby, and H. Münzel, *Francium: Gmelin Handbook of Inorganic Chemistry*, 8th ed., System No. 25a (Springer, Berlin, 1983).
63. A. K. McMahan, Physica **139&140B**, 31 (1986).
64. J. Meyer-ter-Vehn and W. Zittel, Phys. Rev. B **37**, 8674 (1988).
65. V. G. Vaks and A. V. Trefilov, Sov. Phys. Solid State **19**, 139 (1977) [Fiz. Tverd. Tela **19**, 244 (1977)].

CHAPTER 6
The Group II Elements (The Alkaline-Earth Metals)

6.1 Introduction

The alkaline-earth metals are similar to the alkali metals in their relatively large volumes and compressibilities and their complex phase diagrams which are strongly influenced by nearby empty *d* bands. However, the divalency of the alkaline earths leads to higher bulk moduli and melting points, with the result that the alkaline-earth phase diagrams are not as thoroughly explored as for the alkalis.

6.2 Beryllium

At RTP, Be is hcp. At RP only about 15 K before melting, Be transforms to bcc. The hcp-bcc transition is accompanied by a 4% increase in density, implying that the slope of the phase boundary is negative[1,2]. The hcp-bcc boundary has been measured to 6.0 GPa[3]. The trend of this curve suggests that there will be an hcp-to-bcc transition at RT and very high pressure.

There has been a serious effort to locate the expected transition at low temperature. Resistance measurements with the DAC show no evidence of the RT transition up to 40 GPa[4]. Careful measurement of release-wave profiles up to 35 GPa in shocked Be also show no evidence of a transition[5]. DAC XRD studies indicate a transition from hcp to a distorted hcp structure with twice as many atoms per unit cell in the range $8.6 < P < 14.5$ GPa[6]. No further changes are seen up to 28 GPa.

The unusual axial ratio ($c/a = 1.56$) and Poisson ratio ($\sigma = 0.04$) of Be suggest strong deviations from simple NFE metallic behavior. This is confirmed by Hartree-Fock and AIP band-structure calculations[7,8] which show a large dip in the density of states at the Fermi energy. The calculations accurately reproduce the anomalous properties of Be, which can be explained in terms of a large spatially anisotropic *p*-electron component in the valence band.

LMTO[9], AIP[10], and ASW[11] calculations on the static hcp, fcc, and bcc structures show that hcp is the most stable phase at RPT, as observed, but that bcc (LMTO, ASW) or fcc (AIP) becomes more stable at pressures above 100 GPa. The bcc and fcc phases are very close in energy. Because the Debye temperature is large in Be, the zero-point energy makes an important contribution to the phase-transition pressure. In bcc Be, the phonons are softer than in hcp, and this stabilizes the bcc phase at high temperatures[10]. Estimates of the Debye temperatures for the various phases from AIP elastic-constant calculations give good agreement with the RP hcp-bcc transition at 1530 K, and predict the RT transition at ca. 100 GPa[10]. These results suggest that experimental searches for the RT transition should be pushed to higher pressure.

The melting curve has been measured to 6.0 GPa[3]. The Be phase diagram is shown in Fig. 6.1.

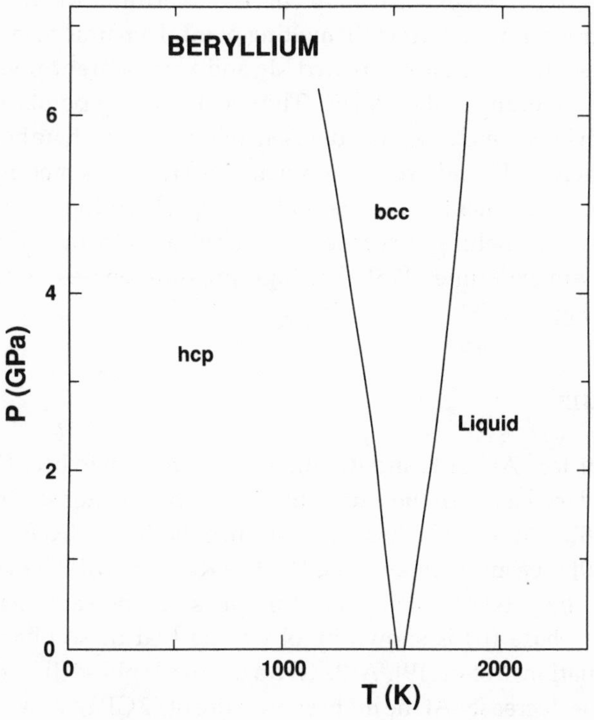

Fig. 6.1 The phase diagram of beryllium.

6.3 Magnesium

At RTP, Mg is hcp. Compression at RT in a DAC to ca. 50 GPa shows a transition to bcc[12]. No further transitions are seen to 58 GPa.

LMTO and GPT calculations for the 0 K static lattice predict the sequence hcp→bcc→fcc[13]. LMTO predicts the two transitions at 57 GPa and 180 GPa, while GPT theory predicts them at 50 and 790 GPa. AIP calculations predict the hcp-bcc transition at 60 GPa[14]. The accurate theoretical prediction of the hcp-bcc transition, carried out before the experimental work, provides strong justification for the validity and utility of the theory. It is interesting that the predicted pressures are strongly dependent on the accurate characterization of the empty *3d* band, which hybridizes with the valence band. Lattice-dynamics calculations using a NFE pseudopotential also show an hcp-bcc phase transition at 85 GPa and 0 K[15]. The theoretical transition pressure drops rapidly with increasing temperature and intersects the melting curve at ca. 10 GPa, giving the Mg phase diagram an appearance similar to that of Be.

The melting curve of Mg has been measured statically to 5.0 GPa[16], as shown in Fig. 6.2. The experimental curve is in error at RP and has been shifted to agree with the correct RP melting point. Dynamic measurements on the interface between shock-loaded Mg and a transparent anvil material have revealed melting of the Mg[17]. The shock-melting points agree with a Lindemann Law which uses a Grüneisen model intermediate between the Dugdale-MacDonald and Free-Volume models[17]. This melting curve is also in good agreement with the static data. Pseudopotential model calculations of the melting curve yield a theoretical curve in slightly poorer agreement with experiment[15]. The high-pressure phase diagram of Mg is shown in Fig. 6.3.

6.4 Calcium

At RTP, Ca is fcc. At 721 K and RP, there is a transition to bcc. The fcc-bcc phase boundary has been measured to 3.5 GPa by a piston-cylinder DTA technique[18]. At RT, Ca has been studied both by XRD[19] and by resistance[20] measurements to 44 GPa. The RT DAC XRD work shows a transition to bcc at 19.5 GPa. That this phase is the same as the high temperature phase (II) is shown by observing that these phases have a common equation of state[19]. At 32 GPa a new sc(1) phase (III) occurs with an 8% volume decrease. At the highest pressure of 42 GPa, weak diffraction lines appear, indicating a new, unidentified phase (IV).

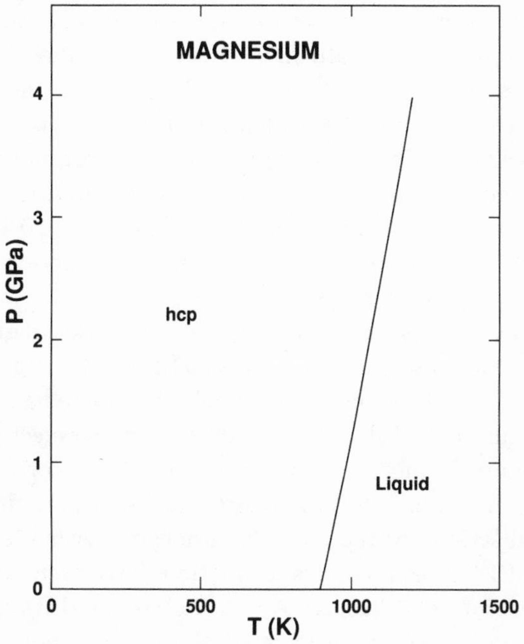

Fig. 6.2 The phase diagram of magnesium at low pressure.

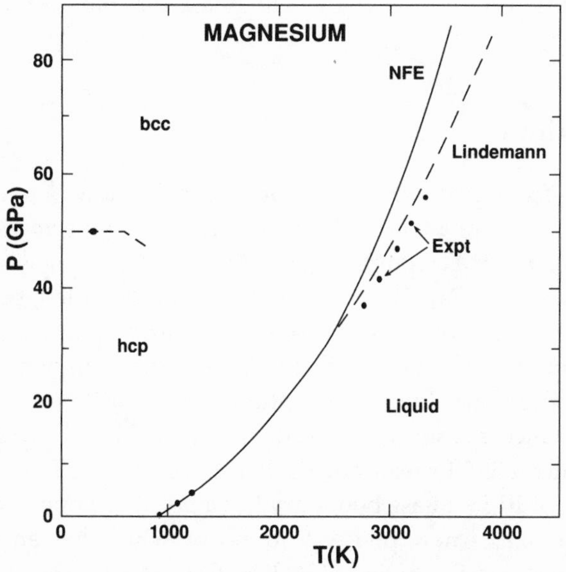

Fig. 6.3 The phase diagram of magnesium at high pressure.

The resistance measurements were carried out in a sintered-diamond-tipped opposed-anvil apparatus[20]. The resistance shows complex behavior, with a maximum at about 18 GPa, a subsequent minimum at 25–30 GPa, a sharp rise with a discontinuity at 36 GPa, and a second maximum at 42 GPa. The two maxima and the discontinuity correlate well with the phase changes observed in the XRD data. In addition, the temperature was varied over the range 2–300 K. The change in position of the main features of the resistance curve with temperature shows that the phase boundaries have positive slope.

As with the heavier alkali metals, the Ca phase diagram is strongly influenced by the presence of the empty 3*d* band. Hybridization effects are already significant at RP, and Ca cannot be considered a good NFE metal. LMTO calculations[21,22] find about 0.6 *d*-character electrons per atom at RP, and this number increases with pressure. LMTO correctly predicts the fcc-bcc phase transition at 21 GPa, but calculations are difficult for open structures like sc(1), and predictions have not yet been made for the bcc-sc(1) transition. GPT lattice-dynamics calculations have been carried out for the fcc-to-bcc transition at RP[23,24]. A transition is found at approximately the right temperature, but the calculated slope is in poor agreement with experiment. As with the 19.5 GPa transition, this one is dependent on *d*-band effects.

The melting curve has been measured to 4.0 GPa[18]. The melting curve is not well predicted by GPT[25]. The phase diagram of Ca is shown in Fig. 6.4.

6.5 Strontium

At RTP, Sr is fcc. RT compression of Sr to 3.5 GPa shows a transition to the bcc phase[18]. The phase boundary has a negative slope and intersects the $P = 0$ axis at 830 K. DAC compression of Sr at RT shows a new structure (III) appearing near 26 GPa[19]. This structure is unidentified, but may be an orthorhombic distortion of the sc(1) Ca III structure. A new structure, Sr IV, appears around 35 GPa[19]. The Sr IV lines gradually strengthen with pressure, indicating a sluggish transition. This transition is probably also seen in resistance measurements with an opposed-anvil apparatus[26]. A kink is seen in the 35 GPa resistance isobar at about 200 K. This suggests that, as in Ca, the Sr III-IV phase boundary has a positive slope. At the highest pressure, 46 GPa, a new phase, V, appears, which has an unidentified structure closely similar to that of Ba IV[19]. Except for the kink at 35 GPa, the resistance data do not clearly indicate any higher-pressure transitions.

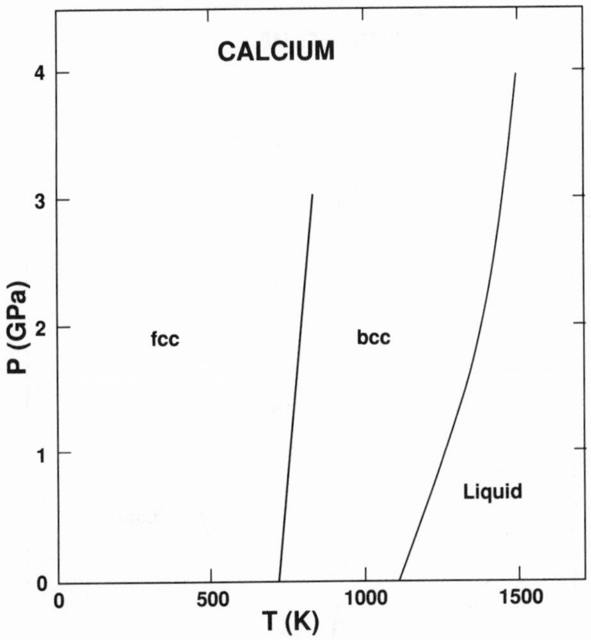

Fig. 6.4 The phase diagram of calcium.

LMTO calculations predict the fcc-bcc transition at about 4.0 GPa, as observed, and a transition to hcp above 50 GPa[21]. These calculations have of course not considered the complex, unknown structures actually observed in Sr above 26 GPa. Several attempts to compute the lower-pressure phase diagram have been made using local-pseudopotential theory and GPT[23,27]. GPT incorrectly predicts a positive dP/dT for the fcc-bcc phase boundary. The local-pseudopotential results are in qualitative agreement with experiment.

The melting curve has been measured to 4.0 GPa[18]. Approximate NFE melting-curve calculations with a modified pseudopotential show adequate agreement with experiment[27]. The phase diagram of Sr is shown in Fig. 6.5.

6.6 Barium

At RTP, Ba is bcc. At 5.5 GPa and RT, the bcc phase transforms to hcp[28]. At 7.5 GPa, another transformation occurs to an hcp-like phase (III)[19]. This phase is probably hcp with a slight distortion, leading to a superlattice with a large unit cell. A further transition at 12.6 GPa yields Ba IV, an

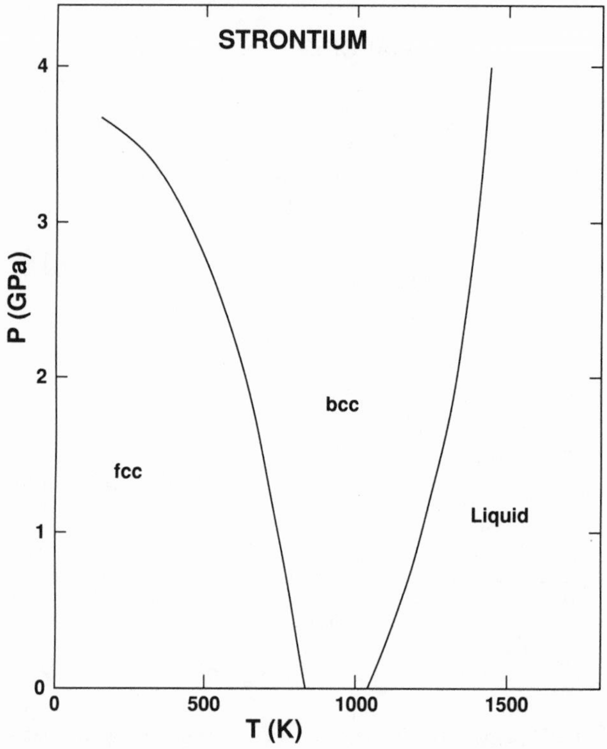

Fig. 6.5 The phase diagram of strontium.

undetermined structure similar to Sr V. No further phase changes are recorded up to 40 GPa[19]. In addition to RT phase transitions, the trajectories of the I-II, II-III, and III-IV phase boundaries have been determined by resistance and DTA measurements[29–32]. There is significant disagreement about the slopes of these boundaries.

Theoretical LMTO calculations predict a bcc-hcp transition at 10.0 GPa[21,22], while AIP calculations predict the transition at 1.1 GPa[33]. GPT lattice-dynamics calculations show that bcc Ba exhibits transition-metal behavior, resulting from the bottom of the 5*d* band falling below the Fermi level[34].

The melting curve of Ba has been measured to 13.0 GPa[29,31,32]. Three maxima are observed. The resemblance to the Cs phase diagram is very strong and there is no doubt that the *s*-*d* transfer is responsible for these features as well as for the solid-solid transitions. The phase diagram of Ba is shown in Fig. 6.6.

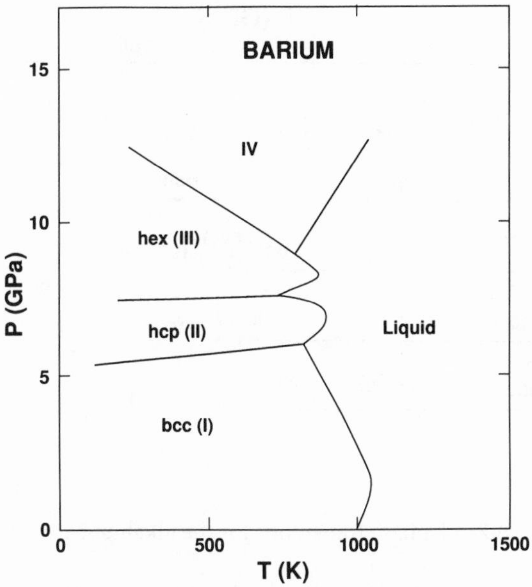

Fig. 6.6 The phase diagram of barium.

6.7 Radium

Ra at RTP is bcc[1]. Because of its strong radioactivity, Ra has not been studied at high pressure. LMTO calculations predict that the bcc structure will remain stable up to 20 GPa[21]. Ra melts at ca. 970 K[35].

6.8 Discussion

The RT phase transitions for the alkaline earths are shown in Fig. 6.7.

Because the alkaline earths are divalent, the filled s^2 band can become insulating if it is separated from the nearby *p* bands. This tendency is seen in the semi-covalent bonding in Be[8] and the semimetallic behavior of fcc Ca and Sr under pressure[20].

As with the alkali metals, the phase diagrams of the heavier alkaline earths reflect the transfer of electrons from *s* to *d* character. The *s-d* transfer can be measured by the effective number of *d* electrons[21,22] or by R_A/R_I[19]. The fcc-to-bcc transitions in Ca and Sr occur at $R_A/R_I \cong 1.95$, and the transformations to the "Ba IV" structures in Sr and Ba occur at $R_A/R_I \cong 1.67$. As seen also in the alkalis, the sequence of structures is not fixed, but depends on the details of the band structure for each metal. The transitions fcc→bcc→complex→Ba IV show the same trend toward lower

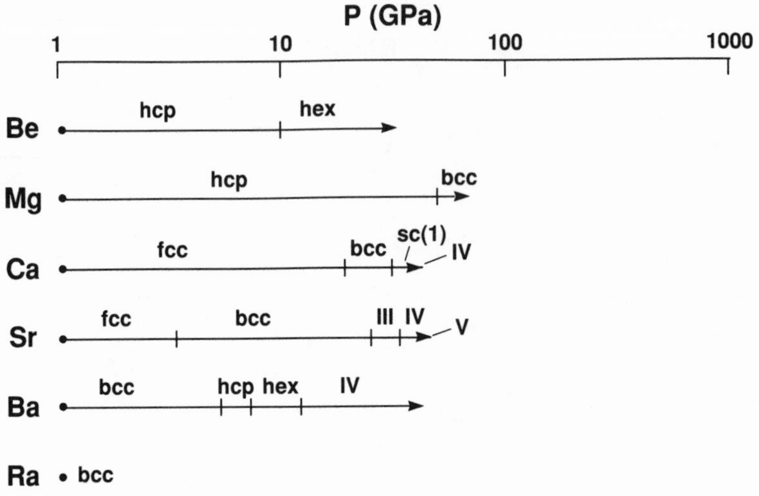

Fig. 6.7 RT phase transitions for the alkaline-earth metals.

pressures with increasing atomic weight as seen in the alkali metals. The increasing area of bcc stability with increasing temperature seen in Be, Mg, Ca, and Sr reflects the general property of entropy stabilization of bcc at high temperatures.

Other elements in the periodic table, namely Eu, Yb, and Es, also exhibit alkaline-earth-like behavior. The nearby partially filled *d* bands in Eu and Yb allow *s-d* transfer under pressure which gives phase diagrams very similar to those of Ba and Sr, respectively. Here is a clear indication of the close correlation of band structure and phase stability.

References

1. J. Donohue, *The Structures of the Elements* (Wiley, New York, 1974) chap. 4.
2. A. J. Martin and A. Moore, J. Less-Comm. Met. **1**, 85 (1959).
3. M. François and M. Contre, *Conférence Internationale sur la Métallurgie du Beryllium* (Presses Universitaires de France, Paris, 1965) p. 201.
4. R. L. Reichlin, Rev. Sci. Instrum. **54**, 1674 (1983).
5. L. C. Chhabildas, J. L. Wise, and J. R. Asay, in *Shock Waves in Condensed Matter—1981*, W. J. Nellis, L. Seaman, and R. A. Graham, eds. (Am. Inst. Physics, New York, 1982) p. 422.
6. L. C. Ming and M. H. Manghnani, J. Phys. F **14**, L1 (1984).
7. R. Dovesi, C. Pisani, F. Ricca, and C. Roetti, Phys. Rev. B **25**, 3731 (1982).
8. M. Y. Chou, P. K. Lam, and M. L. Cohen, Phys. Rev. B **28**, 4179 (1983).
9. A. K. McMahan, in Ref. 5, p. 340.
10. P. K. Lam, M. Y. Chou, and M. L. Cohen, J. Phys. C **17**, 2065 (1984).

11. J. Meyer-ter-Vehn and W. Zittel, Phys. Rev. B **37**, 8674 (1988).
12. H. Olijnyk and W. B. Holzapfel, Phys. Rev. B **31**, 4682 (1985).
13. A. K. McMahan and J. A. Moriarty, Phys. Rev. B **27**, 3235 (1983).
14. R. M. Wentzcovitch and M. L. Cohen, Phys. Rev. B **37**, 5571 (1988).
15. J. L. Pelissier, Phys. Scr. **34**, 838 (1986).
16. G. C. Kennedy and R. C. Newton, in *Solids under Pressure*, W. Paul and D. M. Warschauer, eds. (McGraw-Hill, New York, 1963) p. 163.
17. P. A. Urtiew and R. Grover, J. Appl. Phys. **48**, 1122 (1977).
18. A. Jayaraman, W. Klement, Jr., and G. C. Kennedy, Phys. Rev. **132**, 1620 (1963).
19. H. Olijnyk and W. B. Holzapfel, Phys. Lett. **100A**, 191 (1984).
20. K. J. Dunn and F. P. Bundy, Phys. Rev. B **24**, 1643 (1981).
21. H. L. Skriver, Phys. Rev. Lett. **49**, 1768 (1982).
22. H. L. Skriver, Phys. Rev. B **31**, 1909 (1985).
23. J. A. Moriarty, Phys. Rev. B **8**, 1338 (1973).
24. J. A. Moriarty, Int. J. Quantum Chem. Symp. **17**, 541 (1983).
25. J. A. Moriarty, personal communication.
26. K. J. Dunn and F. P. Bundy, Phys. Rev. B **25**, 194 (1982).
27. J. L. Pelissier, Phys. Lett. **103A**, 345 (1984).
28. J. C. Haygarth, I. C. Getting, and G. C. Kennedy, J. Appl. Phys. **38**, 4557 (1967).
29. A. Jayaraman, W. Klement, Jr., and G. C. Kennedy, Phys. Rev. Lett. **10**, 387 (1963).
30. B. C. Deaton and D. E. Bowen, Appl. Phys. Lett. **4**, 97 (1964).
31. J. P. Bastide and C. Susse, High Temp.-High Press. **2**, 237 (1970).
32. A. Yoneda and S. Endo, J. Appl. Phys. **51**, 3216 (1980).
33. Y. Chen, K. M. Ho, and B. N. Harmon, Phys. Rev. B **37**, 283 (1988).
34. J. A. Moriarty, Phys. Rev. B **34**, 6738 (1986).
35. R. J. Meyer, *Radium und Isotope: Gmelins Handbuch der Anorganischen Chemie* 8th ed., System No. 31 (Verlag Chemie, Berlin, 1928) p. 54.

CHAPTER 7
The Group III Elements

7.1 Introduction

The phase diagrams of the Group III elements have not yet received the kind of attention given to the alkalis and alkaline-earth metals. Because the heavier Group III elements have filled *d* bands, the *s-d* transfer effects seen in Groups I and II are not seen here. However, the group does manifest a tendency toward lattice distortions which are usefully understood in terms of NFE pseudopotential theory.

7.2 Boron

The phase diagram of B is virtually unknown. Numerous complex crystal structures have been discussed in the literature[1,2], but their relative thermodynamic stability is not known. The tendency toward covalency seen in Be is fully developed in B, but with only 3 bonding electrons, B is electron deficient, and it cannot form a simple 3-dimensional network like diamond with a filled valence band. The "solution" to this problem is a covalent semiconductor with lattices containing icosahedral B_{12} units, as shown in Fig. 7.1. The icosahedral atoms are each bonded to 5 neighbors by three-center bonds and the icosahedra are connected to each other by stronger covalent bonds. The result is a rather open lattice which readily accepts metallic atoms into its interstices to produce boron-rich borides. It is possible that some of the structures originally proposed for pure B are in fact those of such borides.

Solidification of liquid B yields the β-rhombohedral structure, rh(105), which has 105 atoms per unit cell and 16 distinct atomic positions[1,2]. This phase is thought to be thermodynamically stable at all temperatures below melting. Other phases which can be prepared readily are α-rhombohedral, rh(12); α-tetragonal, st(50); and β-tetragonal, st(192)[1,2]. The rh(105) phase

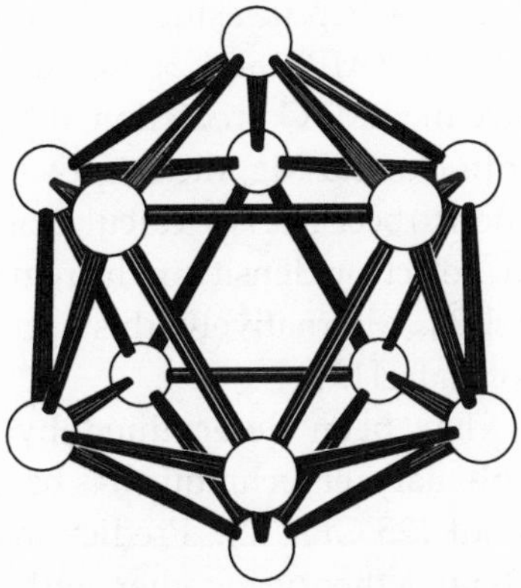

Fig. 7.1 Structure of the boron B_{12} icosahedron. (From H. L. Yakel, AIP Conf. Proc. **140**, 97 [1986]. Redrawn with permission.)

shows anomalous behavior near 180 K, which may indicate a phase change[3].

Static high-pressure resistance measurements up to 25 GPa show no evidence of phase transitions[4]. A new B phase of unknown structure was reported above 10 GPa and 2000 K, but the probability of sample contamination in this work was high[5]. Shock compression to 110 GPa shows no evidence of any phase change[6].

Theoretical work on B has until now been neglected. Hartree-Fock calculations on icosahedral clusters of B atoms clearly show the 3-center bonds[7]. New AIP and LMTO total-energy calculations on B reveal transitions from covalent B to metallic ct(2) at 210 GPa and from ct(2) to fcc at 360 GPa[8]. The insulator-metal transition at 210 GPa is within range of DAC technology.

B expands very slightly upon melting at 2365 K, which implies a melting curve with a positive slope[9]. Since the solid is already a very open structure, the liquid must share its basic structure, possibly as independent B_{12} units[9]. The melting curve has not been directly measured.

7.3 Aluminum

Al is fcc at RTP. DAC compression of Al to 150 GPa at RT shows no change of phase[10,11]. Also, sound-speed measurements on the shock Hugoniot of Al up to the melting point near 125 GPa show no evidence of a phase change in the solid[12].

Recent GPT, LMTO, and AIP total-energy calculations show that the sequence of transitions fcc→hcp→bcc is to be expected at pressures above

100 GPa[13,14]. For the fcc→hcp transition, the predicted pressures are 120 GPa (LMTO), 240 GPa (AIP), and 360 GPa (GPT). Although the transition must lie above the LMTO prediction, it has not yet been located experimentally. As with Na and Mg, the empty 3*d* band is important in determining the transition to bcc. The AIP calculations show clearly that bcc is favored by increased *d*-electron density in the region between atoms and their second neighbors[14]. Alternatively, this appears as a destabilizing dip in the fcc density of states[13].

The melting curve has been determined by static techniques to 6.0 GPa[15]. The optical-analyzer technique has been used to find melting on the Hugoniot at about 125 GPa[12]. Predictions of the melting curve using NFE pseudopotential theory together with lattice dynamics and liquid perturbation theory are in good agreement with experiment over the whole range from 0 to 150 GPa[16,17]. The low- and high-pressure Al phase diagrams are shown in Figs. 7.2 and 7.3.

7.4 Gallium

Solid Ga I at RTP is eco(8)[1]. Each atom has only one nearest neighbor, with six other neighbors somewhat farther away. Below RT, there is a transition at roughly 2 GPa to Ga II, which is bcc(12)[18]. The I-II transition has been measured from the I-II-liquid triple point down to about 100 K[19,20]. Compression at higher temperatures leads to a new phase, III, which is ct(2), isostructural with In[18]. The II-III phase boundary has been measured to 7.5 GPa[19]. There is a II-III transition at RT and ca. 14 GPa which means a negative slope in the II-III phase boundary at high pressure[21]. No new phases are found up to 30 GPa[21].

Theoretical explanations for the rather open Ga structures are based on the physics of trivalent NFE metals[22,23]. The effective pair potential computed from NFE theory has an undulation with a local maximum at a distance close to that of the close-packed nearest neighbors. The lattice can lower its energy by a distortion away from close packing which avoids the potential maximum. This is shown schematically in Fig. 7.4.

The melting curve of Ga has been measured to 7.5 GPa[19]. The initial slope is negative, but becomes positive at the I-II-liquid triple point. The very low melting temperature of Ga may also be explained in terms of the anomalous pair potential in that the distorted lattice has a rather high free energy, and only a small amount of thermal energy is required to stabilize the disordered liquid. The Ga phase diagram is shown in Fig. 7.5.

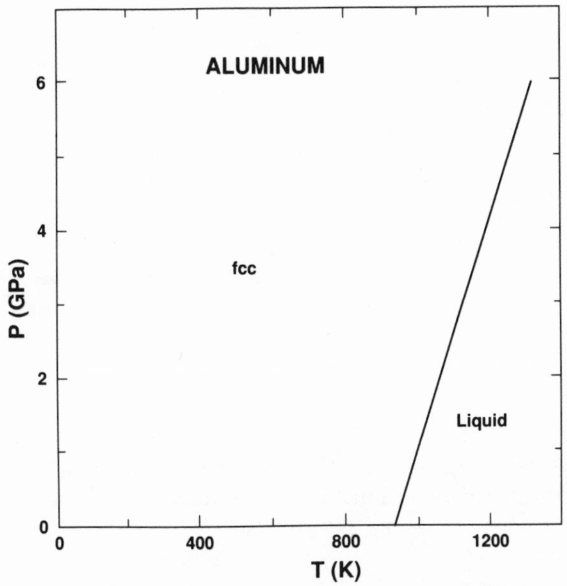

Fig. 7.2 The phase diagram of aluminum at low pressure.

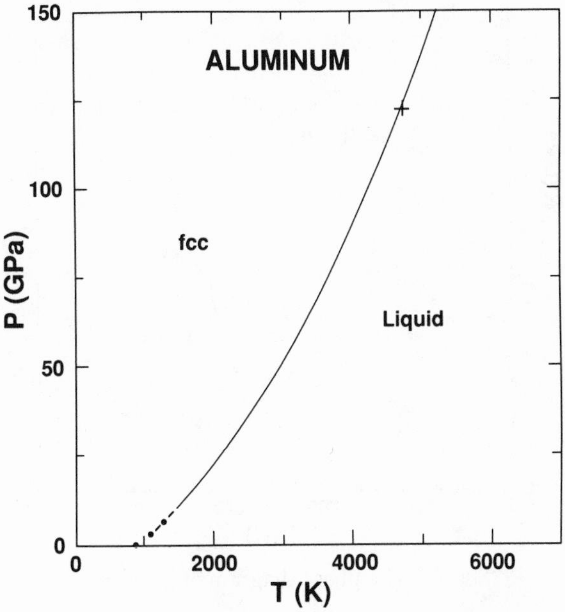

Fig. 7.3 The theoretical melting curve of aluminum at high pressure. The points are low-pressure experiments and the cross is the predicted shock-Hugoniot melting point.

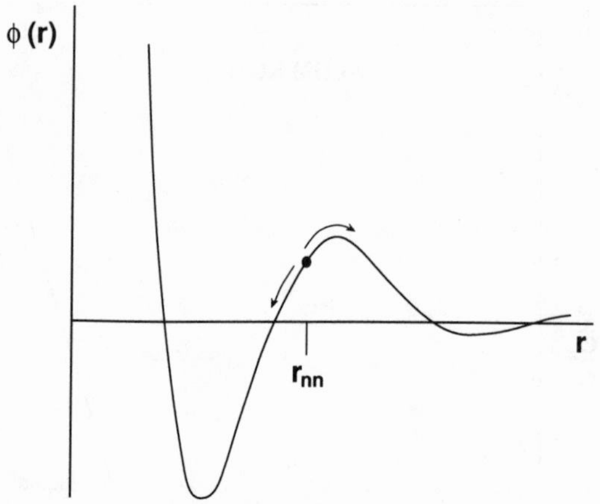

Fig. 7.4 A schematic drawing of the Ga pair potential, showing the direction that nearest-neighbor atoms must move to minimize the lattice energy.

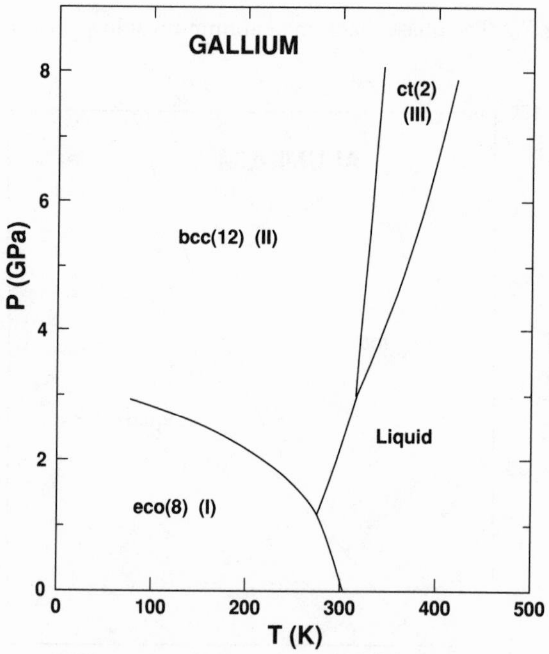

Fig. 7.5 The phase diagram of gallium.

7.5 Indium

At RTP, In is ct(2), isostructural with Ga III. RT compression up to 40 GPa shows no sign of any phase change[21,24,25]. The melting curve has been measured to 7.5 GPa[19]. The In phase diagram is shown in Fig. 7.6.

7.6 Thallium

At RTP, Tl is hcp. At RP and 503 K, there is a transition to bcc, which is stable up to the melting point. At RT and ca. 3.7 GPa, hcp transforms to fcc[19]. RT compression to 38 GPa shows no further change of phase[21]. The three phases meet at a triple point at 3.85 GPa and 388 K[19]. There is disagreement about the sign of the slope of the hcp-fcc phase boundary[19,26,27]. The fcc-bcc phase boundary has been measured to 8.5 GPa[26]. Its slope increases at the highest pressure, making it unlikely that there will be a bcc-fcc-liquid triple point. The melting curve has been measured to 5.0 GPa[19]. The phase diagram of Tl is shown in Fig. 7.7.

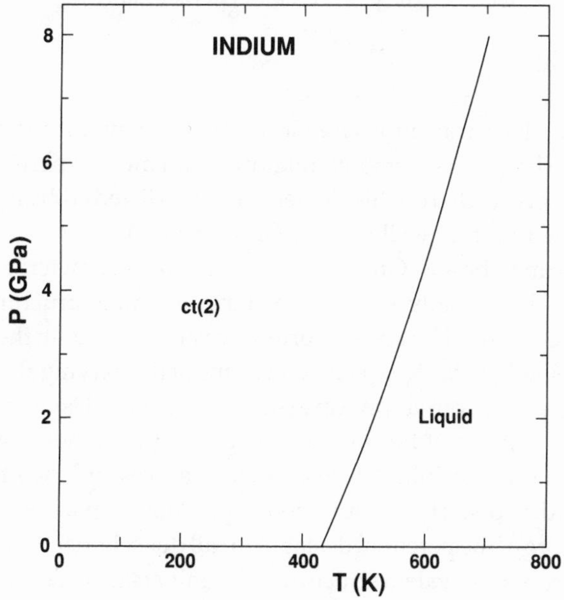

Fig. 7.6 The phase diagram of indium.

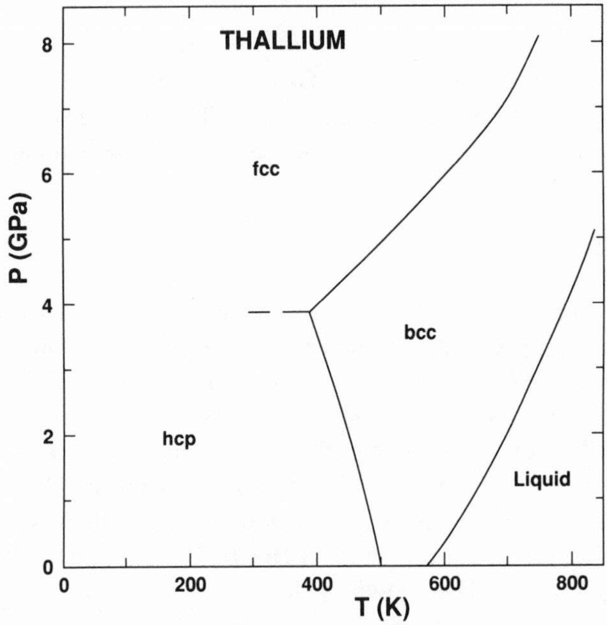

Fig. 7.7 The phase diagram of thallium.

7.7 Discussion

The isothermal phase transitions for Group III are summarized in Fig. 7.8.

The B phase diagram is virtually unknown, and more experimental and theoretical work is needed on this element. The predicted ct(2) and fcc high-pressure phases fit rather well into the Group III pattern.

The metallic members of Group III are fairly well characterized experimentally and Al especially so. The pattern of stable structures in the sequence Al→Ga→In→Tl is cp→distorted→distorted→cp. In the model of Hafner and Heine[23], this is explained in terms of the varying shapes of the pair potentials arising from empty-core pseudopotentials. Using expressions for the elastic constants of the simple lattice structures, they can map the regions of mechanical stability of close-packed lattices and their tetragonal and rhombohedral distortions. A region of parameter space occurs where the repulsive core of the potential destabilizes all densely packed structures and favors more open covalent structures. A "universal phase diagram" for the Group III elements can be constructed from r_c, the pseudopotential-core radius, and r_s, the electron-sphere radius, as shown in Fig. 7.9. Here the predicted stable structures of Al, Ga, In, and Tl, and their variation with pressure, are given. Al is well within the region of close-packed stability. Ga is in the "open structure" region (Ga I) but moves into tetragonal stability (Ga III) under pressure. In favors tetragonal, and Tl has moved back into a region of close-packed stability. The NFE model thus explains the puzzling Group III structure sequence, at least qualitatively.

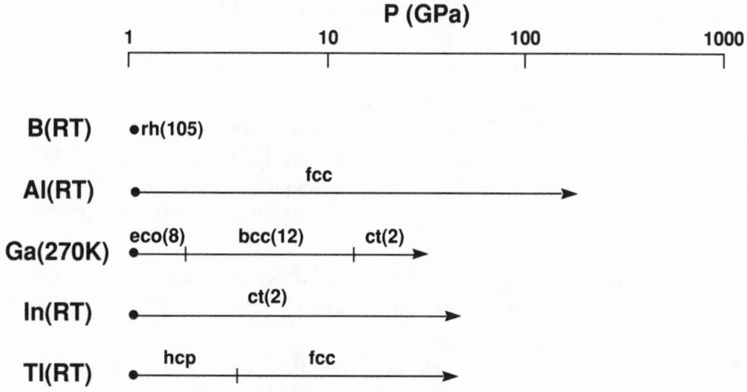

Fig. 7.8 The isothermal phase transitions of the Group III elements.

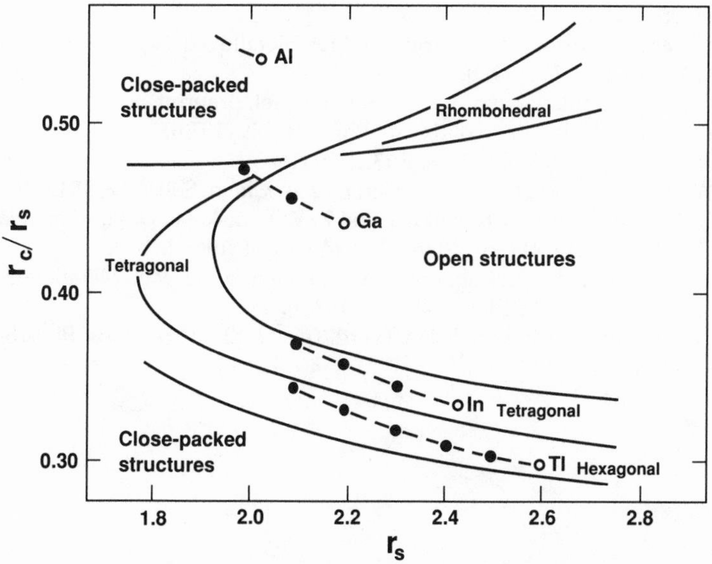

Fig. 7.9 The Group III phase diagram predicted by NFE theory. (From Hafner and Heine[23]. Redrawn with permission.)

References

1. J. Donohue, *The Structures of the Elements* (Wiley, New York, 1974) chap. 5.
2. E. Amberger and W. Stumpf, *Boron: Gmelin Handbook of Inorganic Chemistry*, 8th ed., System No. 13 (Springer, Berlin, 1981).
3. H. Werheit and R. Franz, Phys. Stat. Sol. (b) **125**, 779 (1984).
4. L. F. Vereshchagin, A. A. Semerchan, S. V. Popova, and N. N. Kuzin, Dokl. Akad. Nauk SSSR **145**, 757 (1962) [Sov. Phys. Doklady **7**, 692 (1963)].
5. R. H. Wentorf, Jr., Science **147**, 49 (1965).
6. S. P. Marsh, ed., *LASL Shock Hugoniot Data* (University of California Press, Berkeley, Los Angeles, London, 1980) p. 24.
7. A. D. Zdetsis and D. K. Papademitriou, Phys. Rev. Lett. **60**, 61 (1988).
8. C. Mailhoit, J. B. Grant, and A. K. McMahan, Phys. Rev. B **42**, 9033 (1990).
9. D. V. Khantadze and N. J. Topuridze, J. Less-Common Met. **117**, 105 (1986).
10. L. C. Ming, D. Xiong, and M. H. Manghnani, Physica **139&140B**, 174 (1986).
11. A. Ruoff, personal communication.
12. R. G. McQueen, J. N. Fritz, and C. E. Morris, in *Shock Waves in Condensed Matter—1983*, J. R. Asay, R. A. Graham, and G. K. Straub, eds. (North-Holland, Amsterdam, 1984) p. 94.
13. A. K. McMahan and J. A. Moriarty, Phys. Rev. B **27**, 3235 (1983).
14. P. K. Lam and M. L. Cohen, Phys. Rev. B **27**, 5986 (1983).
15. J. Lees and B. H. J. Williamson, Nature **208**, 278 (1965).
16. J. A. Moriarty, D. A. Young, and M. Ross, Phys. Rev. B **30**, 578 (1984).
17. J. L. Pelissier, Physica **128A**, 363 (1984).
18. L. Bosio, J. Chem. Phys. **68**, 1221 (1978).
19. A. Jayaraman, W. Klement, Jr., R. C. Newton, and G. C. Kennedy, J. Phys. Chem. Solids **24**, 7 (1963).
20. T. M. Turusbekov and E. I. Estrin, Fiz. Met. Metalloved. **52**, 651 (1981) [Phys. Met. Metall. **52**, 178 (1981)].
21. O. Schulte, A. Nikolaenko, and W. B. Holzapfel, preprint.
22. V. Heine and D. Weaire, Solid State Phys. **24**, 249 (1970).
23. J. Hafner and V. Heine, J. Phys. F **13**, 2479 (1983).
24. R. W. Vaughan and H. G. Drickamer, J. Phys. Chem. Solids **26**, 1549 (1965).
25. L. F. Vereshchagin, S. S. Kabalkina, and Z. V. Troitskaya, Dokl. Akad. Nauk. SSSR **158**, 1061 (1964) [Sov. Phys. Doklady **9**, 894 (1965)].
26. T. Ye. Antonova, I. T. Belash, and S. A. Ivakhnenko, Fiz. Met. Metalloved. **49**, 438 (1980) [Phys. Met. Metall. **49**, 196 (1980)].
27. M. A. Il'ina and E. S. Itskevich, Fiz. Tverd. Tela **12**, 1240 (1970) [Sov. Phys. Solid State **12**, 965 (1970)].

CHAPTER 8
The Group IV Elements

8.1 Introduction

The Group IV elements, especially C, Si, and Ge, are important industrial and electronic materials, and this has greatly stimulated research on their phase behavior. The exploration of the graphite-diamond phase boundary, for example, has been driven by the prospect of manufacturing synthetic diamonds.

In recent years DAC experiments together with the advance of theoretical methods have greatly improved our understanding of the Group IV phase diagrams. These diagrams provide a paradigm for the rule of increasing coordination number and the increasing importance of the electrostatic energy with compression.

8.2 Carbon

Because of the enormous technological importance of carbon in the forms of graphite and diamond, the carbon phase diagram has been studied with great persistence and ingenuity for many decades[1–3]. The combination of high pressures and high temperatures required for this study has led to the construction of many unique pieces of experimental apparatus.

Carbon at RTP is graphite, hex(4), shown in Fig. 8.1. In this structure the hexagonal layer planes are benzeneoid and strongly bonded, while the interplanar bonding is much weaker, involving the overlap of π orbitals. The layer planes are stacked in the order ABAB.... There have been arguments for a possible nonhexagonal stacking symmetry for graphite, but the consensus favors the hexagonal form[4].

Graphite which has been recovered after heating above 2600 K shows XRD patterns suggesting that carbon takes on a polyacetylenic structure consisting of straight and kinked chains of double- or triple-bonded

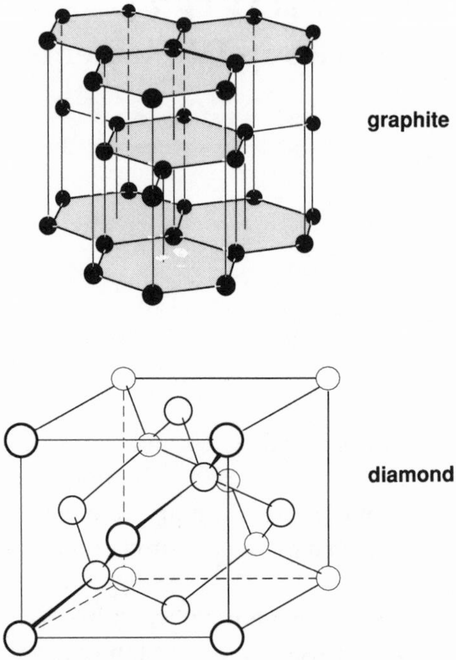

Fig. 8.1 Graphite and diamond crystal structures. (From N. W. Ashcroft and N. D. Mermin, *Solid State Physics* (Saunders College, Philadelphia, 1976) and from Donohue[4] © 1974 John Wiley & Sons, Inc. Redrawn with permission.)

atoms[5,6]. There are 6 of these "carbyne" phases identified, and it has been suggested that they exist in a sequence of short temperature intervals between 2600 K and the solid-liquid-gas triple point[5]. At present there is no direct evidence for the stability of these phases at high temperatures.

The high-pressure phase of C is diamond, fcc(8), shown in Fig. 8.1. The graphite-diamond phase boundary has been intensively studied, and is now accurately known[7–9]. The manufacture of artificial diamonds is based on this knowledge. For $T < 1200$ K, the boundary is determined from measured graphite and diamond thermodynamic data. For 1200 K $< T <$ 2800 K, the boundary is determined from direct conversion using transition-metal catalysts. For $T > 2800$ K the boundary is inferred indirectly from uncatalyzed conversion experiments.

Because the covalent bonds in graphite and diamond are so different, direct conversion of one to the other is very difficult. Graphite will not convert to diamond at RT even if compressed far beyond the transition pressure. It will convert if shocked to 30–50 GPa because the shock heating will permit breakage of the graphite bonds[10,11]. Static compression of graphite above the phase boundary at $T > 3000$ K will also allow diamond formation[8].

In the 1970s experiments and theoretical studies suggested that diamond transforms to a metallic phase at 100–200 GPa[12,13]. However, more recent work shows that this is incorrect. There is no conclusive evidence for metallization from shock-wave studies of diamond[14,15]. Also, the lack of any visible change in DAC surfaces exposed to pressures above 300 GPa indicates that no phase change is occurring in diamond at these pressures.

Extensive total-energy calculations have been performed on C in many different crystal structures. Graphite has been studied with the AIP and FPLAPW methods[16,17]. The theoretical equilibrium lattice parameters at $P = 0$ are in accurate agreement with experiment. Theoretical bulk moduli and elastic constants are in less good agreement with experiments, but experimental uncertainties and the effects of phonons make the comparisons inexact. The calculations on graphite are very demanding because of the very different interplanar and intraplanar bonding. It is clear that the local-density approximation accurately represents the ground state of solid carbon.

AIP and LMTO calculations[18–23] on diamond give excellent agreement with the experimental equation of state. The theoretical graphite and diamond energies are so nearly equal[16] that the 0 K transition pressure is lost in the numerical noise. This is consistent with the low (ca. 1.5 GPa) transition pressure at 0 K. Semiempirical models such as the Debye model, when fitted to the graphite and diamond equations of state, are able to reproduce accurately the graphite-diamond phase boundary[24,25]. The accurate total-energy calculations on graphite and diamond give us confidence that the theory may be extended to higher pressures, and that they will give good predictions of further phase transitions in carbon.

A search for stable high-pressure C phases has shown that the BC-8 (bcc(8)) phase replaces diamond near 1.1 TPa[19,20,22,23]. This phase is a semimetallic distorted-diamond structure. This transition puts an upper bound on the pressure achievable in the DAC, since a phase transition will likely lead to the mechanical failure of the diamond anvils. There could, of course, exist another unknown phase with a still lower transition pressure. The metallic six-coordinated sc(1) and ct(4) (β-Sn) structures have been considered as likely high-pressure phases of C, but these are mechanically unstable with respect to diamond[23]. Transitions to the bcc and close-packed lattices are expected in the 10 TPa range. The theoretical total-energy curves are shown in Fig. 8.2.

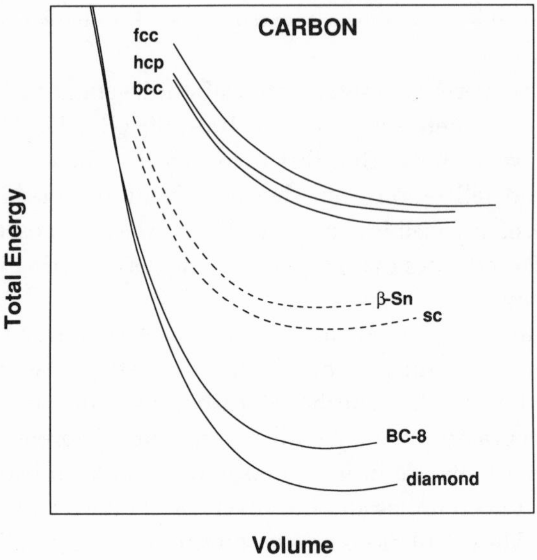

Fig. 8.2 Total energies of carbon phases calculated from AIP theory. (From Fahy and Louie[23]. Redrawn with permission.)

The unusual stability and high density of diamond C are due to the absence of *p*-electron states in the $1s^2$ atomic core[18]. This allows the strongly *p*-character sp^3 bonding electrons in diamond to be held close to the nucleus. In fact, diamond is denser than the predicted zero-pressure states of the close-packed metallic phases, as seen in Fig. 8.2.

Because of the very high value of the graphite melting temperature, there have been many conflicting measurements of the solid-liquid-vapor triple point. These have varied over the range 3800 K to 5100 K[26,27]. There is wide agreement, however, on the triple-point pressure at 11.0 MPa. Recent isobaric heating experiments on graphite show a melting temperature of 4500 ± 100 K[28].

There have been two attempts to measure the complete melting curve of graphite and both show a maximum in the melting temperature at ca. 6.0 GPa[29,30]. The appearance of diamond obtained from melting graphite at ca. 12.0 GPa suggests that this is the pressure of the graphite-diamond-liquid triple point[7].

The question whether diamond melts directly to a liquid phase or first transforms to a solid metallic phase has been raised in a theoretical paper[31]. There is now evidence from a DAC experiment that diamond does indeed melt, although the equilibrium phase boundary has not yet been measured[32,33]. It has frequently been suggested that the carbon phase diagram is analogous to that of Si in having a negative dP/dT for the diamond-liquid phase boundary[1]. However, a recent shock-wave experiment with the optical-analyzer technique found no discontinuity in the sound speed of diamond in the range 100–200 GPa and 5000–6000 K, which is where melting would be expected according to the analogy[34]. Hence either the melting transition has a very small change in sound speed not detected by the experiment, or the melting curve has a positive slope. Additional evidence for a positive slope comes from resistance heating of carbon samples in a high-pressure cell[35]. These experiments show that at pressures above the graphite-diamond transition, the enthalpy, and by inference the temperature, at melt increases with increasing pressure. The low-pressure experimental phase diagram of C is shown in Fig. 8.3.

The structure of liquid C is unknown. Recent quantum-molecular-dynamics calculations predict that the liquid at 5000 K and low pressure has mainly 2- and 3-fold atomic coordination[36]. The 2-fold coordinated atoms form chains similar to the linear C_n molecules found in the vapor[37]. Calculations on liquid C at 9000 K and pressures above 100 GPa indicate mainly 4-fold coordination, as in diamond[38]. Thus according to theory,

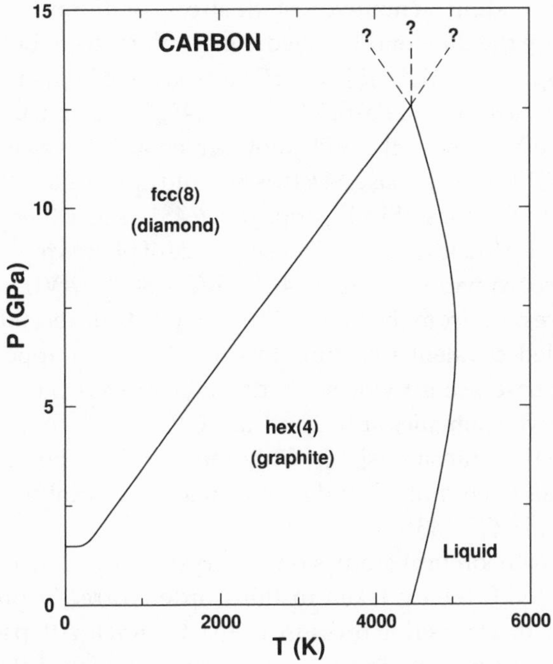

Fig. 8.3 The low-pressure carbon phase diagram.

there is a smooth change in liquid structure from graphite-like to diamond-like as pressure is increased.

There have been several attempts to model the melting curve of graphite and diamond. One approach is to consider the liquid as a pseudobinary mixture of graphite-like and diamond-like states, whose composition changes smoothly with increase in pressure[24,39]. This approach yields a melting curve maximum, as seen in graphite. One recent theoretical paper suggested that the low- and high-pressure liquid states are separated by a first-order phase transition[40]. Another approach is to use free-energy models and to compute the phase boundary in the standard manner[25]. Quantum-molecular-dynamics calculations of P and T along a high-density isochore clearly and convincingly indicate the melting of diamond with a positive dP/dT[38].

8.3 Silicon

Si at RTP is in the cubic diamond fcc(8) semiconductor phase. RT compression to 11.3 GPa under nearly hydrostatic stress causes a transition to the metallic β-Sn ct(4) phase[41,42]. The transition pressure varies from 9.0 to 13.0 GPa, depending on nonhydrostatic stress conditions[41,43–45]. In the ct(4) structure the diamond tetrahedra can still be seen, but are flattened. Further compression[41,42,45] to 13.2 GPa produces Si V, a simple-hexagonal phase, hex(1). At ca. 37 GPa, Si VI appears[45,46]. The structure of Si VI is similar to the X-phase of the BiPb alloy, which may have an orthorhombic structure[47]. The next phase, Si VII, is hcp and appears at 42 GPa[42,45,46]. At 79 GPa, the fcc phase, Si VIII, appears[46,48]. No further phase changes are observed up to 248 GPa[46]. This sequence of phases represents a steady increase in coordination number: 4(I)→6(II)→8(V)→?(VI)→12(VII,VIII).

Decompression from the Si II ct(4) phase yields the bcc(8) BC-8 tetrahedrally bonded covalent structure, Si III[41,49]. The appearance of this metastable phase is sensitive to nonhydrostatic stresses. Shock-wave studies on Si show two anomalies at 10.2 and 13.5 GPa, which could be ascribed to the I-II and II-V transitions[50]. Only the I-II phase boundary has been experimentally determined, and it ends in a triple point with the liquid at 1080 K and 15.0 GPa[43].

Numerous theoretical studies on the crystal structures of Si have been conducted. NFE theory taken to third order correctly predicts the I-II transition, with a transition pressure of 15.0 GPa[51]. AIP pseudopotential calculations have been the most successful in treating the I-II transition, and separate calculations give static lattice transition pressures of 7.0 and

9.3 GPa[20,52]. The II-V transition pressure is predicted to occur at 12.0 and 14.3 GPa[52,53]. A transition to hcp (VII) is predicted at 41.0 GPa[52]. The calculations accurately predict the *c/a* ratios of the II and V phases observed by XRD. LMTO, GPT, LCAO, and AIP calculations predict the hcp-fcc transition at 76, 80, 88, and 116 GPa, respectively[52,54,55]. The overall agreement between theory and experiment is very good.

At still higher pressures, above 100 GPa, LMTO and GPT theory predict an fcc-to-bcc transition[52,54]. The theoretical Si total-energy curves are shown in Fig. 8.4.

The calculations also provide useful insight into the bonding in the various structures of Si. The sequence of transitions can be viewed as a shifting balance between exchange-correlation energy favoring open structures and the Madelung energy favoring close-packed structures. The calculations show a high energy barrier (ca. 5 eV) between the I and II phases, which is consistent with the observed sluggishness of the transition[56]. Calculations on graphitic hex(4) and BC-8 bcc(8) Si clearly

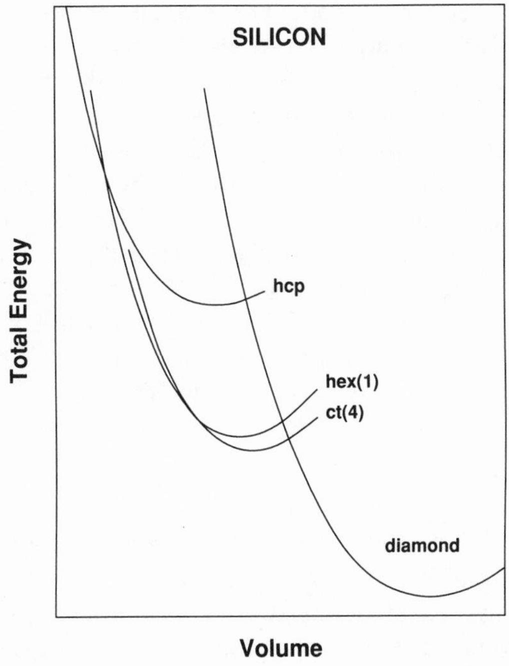

Fig. 8.4 Total energies of silicon structures calculated from AIP theory. (From K. J. Chang and M. L. Cohen, Phys. Rev. B **30**, 5396 (1984). Redrawn with permission.)

reveal the metastability of these phases[16,22]. In addition, the calculations show that the hex(1) V phase has strong interplanar bonding, which is just the opposite of graphite[52]. Calculation of phonon frequencies in Si V gives a low-frequency interlayer acoustic phonon, which suggests a strong temperature effect on the Si II-V phase boundary. At higher pressures, the Si V phase shows a soft phonon which destabilizes it with respect to Si VII[52].

The melting curve of Si has been measured to 20 GPa[43]. The Si I-liquid boundary has a negative slope, which follows from the high density of the more nearly close-packed metallic state of the liquid. Above the I-II-liquid triple point, the melting slope is positive. There is at present no evidence of a II-V-liquid triple point on the experimental melting curve.

First-principles calculations of the melting curve are difficult because there is as yet no accurate representation of the covalent sp^3 bonds which can be used in a lattice-dynamics calculation. Perhaps the most straightforward approach is through third-order NFE perturbation theory for the solid and the use of the Lindemann law to compute the melting curve[57]. This theory obtains a negative melting slope, but not with quantitative accuracy. A more empirical model which treats the bonds as spherical distributions of negative charge has been used, together with a model of the liquid as a NFE metal, to obtain a melting curve with a negative slope in agreement with experiment[58]. However, several parameters were fitted to experimental data in order to achieve this agreement. Liquid metallic Si has a rather low coordination number, unlike the melts of close-packed metals. How a liquid can retain a residual covalency has been dealt with theoretically in the quantum-molecular-dynamics method of Car and Parrinello[59]. The Si ions are free to diffuse, but if they approach one another more closely than a critical distance, a covalent bond is formed. These bonds are constantly being created and destroyed as the ions move around. The phase diagram of Si is shown in Fig. 8.5.

8.4 Germanium

At RTP, Ge is cubic diamond, fcc(8). Compression at RT to 10.6 GPa produces the β-Sn metallic phase, ct(4)[60]. As with Si, this transition is very sensitive to nonhydrostatic shear stresses, and much lower transition pressures are obtained in the presence of shear[60]. Further compression to 75 GPa yields a hex(1) simple-hexagonal phase[61]. At 102 GPa, the hex(1) structure transforms to a dhcp hex(4) structure. There are no further transformations up to 125 GPa[61]. Decompression of Ge II yields the

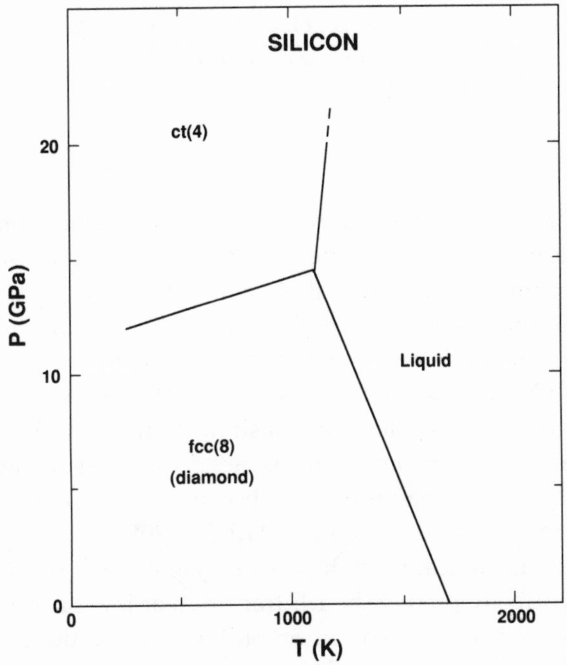

Fig. 8.5 The phase diagram of silicon.

metastable st(12) phase, Ge III[61]. Shock-wave experiments show a likely phase transition, probably I-II, at ca. 12 GPa[62]. The I-II phase boundary has been measured to the I-II-liquid triple point[41].

As with Si, third-order NFE theory predicts the I-II phase transition[51]. AIP calculations show good agreement with experiment for the I-II transition volumes and pressure[63]. The theory also predicts the β-Sn to hex(1) transition at 84 GPa. It does not find dhcp to be stable; rather it predicts a transition from hex(1) to hcp at 105 GPa. The overall agreement between theory and experiment is good.

That Ge has higher transition pressures than Si is in sharp disagreement with the usual trend toward lower pressures with increasing atomic number. The theoretical explanation is that the $3d$ core in Ge repels d-like valence electrons, which in turn raises the energy and pressure of the metallic phases with these electrons[63].

The Ge melting curve has been measured to 20 GPa[43]. The original measurements showed linear P-T trajectories for the Ge I and II melting curves, but more recent measurements indicate some curvature[64]. The I-II-liquid triple point is estimated to occur at 10.5 GPa and 770 K.

Theoretical calculations, as for Si, clearly show the negative slope of the melting curve[57,58]. The Ge phase diagram is shown in Fig. 8.6.

8.5 Tin

Below 291 K at RP, Sn is in the semiconducting I (or α) cubic diamond phase, fcc(8). Above 291 K, Sn exists in the II (or β) ct(4) phase. The I-II boundary has been determined down to 78 K, where the transition pressure is 0.9 GPa[65]. Compression at RT to 9.2 GPa produces Sn III, ct(8)[66]. The II-III phase boundary decreases with increasing temperature, ending in a triple point with the liquid at 2.9 GPa and 581 K[67]. Further compression at RT above 40 GPa yields a sluggish transition to Sn IV, which is bcc[68–70]. The ct(8) structure continuously approaches bcc as pressure increases, but apparently changes discontinuously to bcc in a first-order transition. No further phase changes are found up to 120 GPa[70].

Rather little theoretical work has been done on Sn. Third-order NFE calculations correctly predict the I-II transition at low pressure[51]. AIP calculations have been carried out on Sn I and II, but these calculations

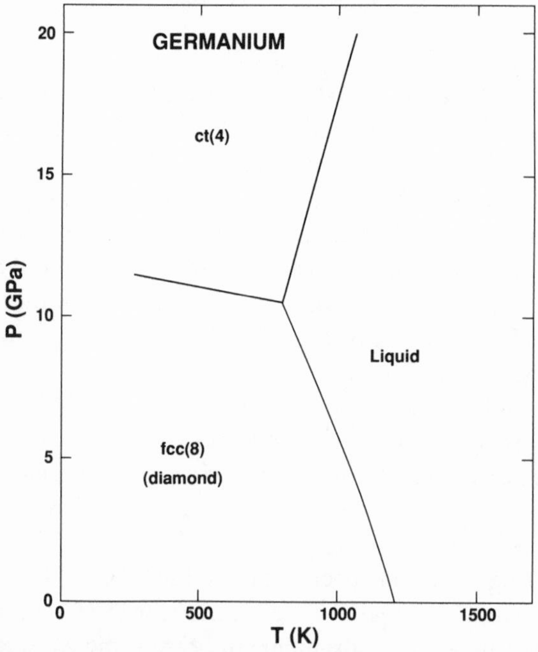

Fig. 8.6 The phase diagram of germanium.

cannot be used to predict the transition because of the exceedingly small I-II energy difference[71].

The melting curve of Sn has been measured to 5.0 GPa[67,72]. The phase diagram of Sn is shown in Fig. 8.7.

8.6 Lead

At RTP, Pb is fcc. Compression at RT to 13.7 GPa yields a transition to hcp with a small volume change[73,74]. DAC XRD measurements show a new bcc phase appearing at 109 GPa[75]. No further change is found up to 280 GPa[75].

LMTO calculations in Pb reveal a strong relativistic "dehybridization" effect, in which the 6*s* energy drops so far below the 6*p* energy that sp^3 hybridization is inhibited[76]. This raises the energy of the diamond structure with respect to fcc. Also of interest is the effect of including spin-orbit relativistic terms as well as scalar relativistic terms. The inclusion of spin-orbit increases the predicted fcc-hcp transition from ca. 0 to 10–15 GPa,

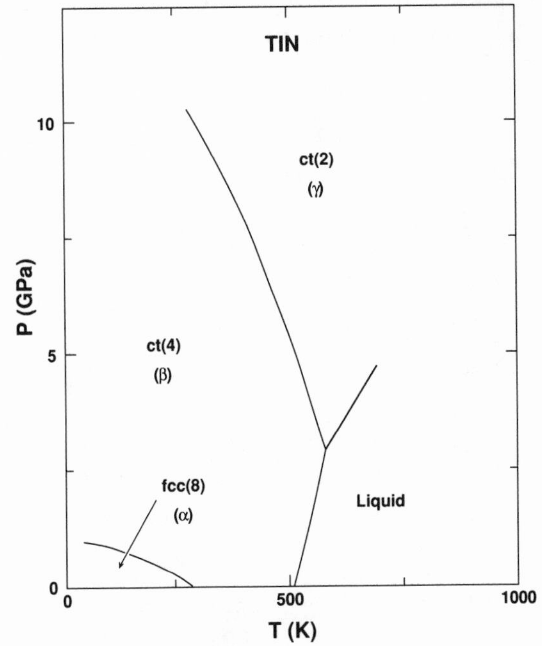

Fig. 8.7 The phase diagram of tin.

in good agreement with experiment[76]. It is evident that without relativistic effects, Pb would resemble Sn, with a stable diamond phase.

The melting curve has been measured to 9.0 GPa in piston-cylinder and opposed-anvil apparatuses[77,78]. Recent DAC experiments with laser heating of the sample have extended the melting curve of Pb to 100 GPa[79]. This measured curve is found to be in good agreement with melting on the shock Hugoniot, which begins at ca. 50 GPa[80,81], and with the prediction of the Lindemann law. NFE pseudopotential calculations are in poor agreement with DAC melting data[82]. The phase diagram of Pb is shown in Fig. 8.8.

8.7 Discussion

The Group IV elements were the first to show clearly the downward trend of phase-transition pressures with increasing atomic number, as indicated in Fig. 8.9. Many authors confidently constructed phase diagrams based on a "corresponding-states" argument which assumed that the same sequence of phases would occur in each member of the group.

With the advance of experiment and theory, it is now clear that the corresponding-states argument is incorrect. Outstanding advances in DAC technology have revealed the structure sequences of Si and Ge to above

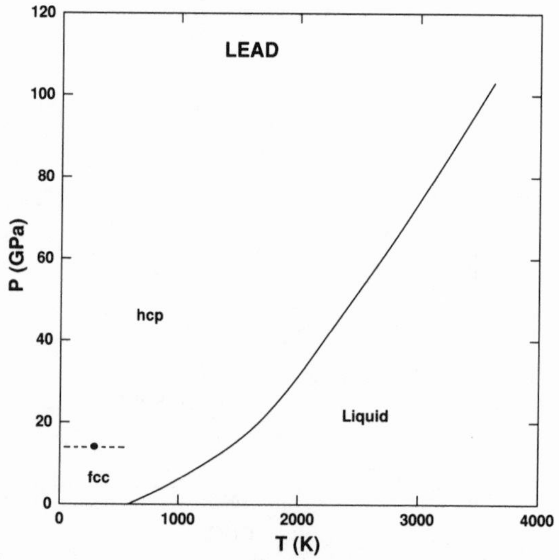

Fig. 8.8 The phase diagram of lead.

100 GPa. There are indeed overall trends, such as the increase in coordination number with pressure, but the structure sequence is modulated by the changes in atomic structure in the Group. A more precise understanding of the Group IV phase behavior is gained by performing total-energy calculations with the local-density approximation.

The calculations have been very successful in predicting the behavior of the Group IV elements under pressure. This success includes the equation of state of the 0 K lattices, the volumes and pressures of the phase transitions, the bonding characteristics of the graphite, diamond, and metallic phases, and even phonon frequencies. It appears that the remaining errors in the theoretical calculations are primarily due to the local-density approximation. The excellent results for measured properties means that we can have high confidence in the predictions for properties at pressures beyond the range of current DAC experiments.

The theory shows that covalency persists even in the intermediate "metallic" phases of Si and Ge, but is finally destroyed in the fcc crystal structure, which shows a NFE-like electron density of states[83]. This process of metallization results from a continuously changing balance between the Madelung, electron exchange-correlation, and electron kinetic energies. The problem for our intuitive understanding of the sequence of phases is that the sequence depends on the details of the band structure. Thus diamond C is unusually stable because of the absence of *p* electrons in the atomic core; Ge has anomalously high transition pressures because of the presence of 3*d* core electrons; and the energies of several phases in Pb are

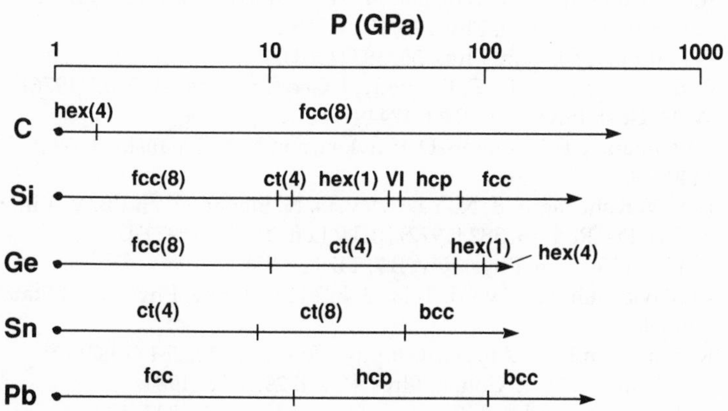

Fig. 8.9 RT phase transitions for the Group IV elements.

shifted because of relativistic terms. These facts were revealed by the calculations and were not obvious to theoreticians before.

Theoretical work on solid-solid and solid-liquid phase boundaries at high temperatures is rather less accurate than the 0 K total-energy calculations, and obviously more work is needed to understand the full phase diagrams of the Group IV elements. Of especial interest is the residual covalency in the liquid states of C, Si, and Ge, and the theoretical models needed to understand this.

For Further Reading

P. Gustafson, "An Evaluation of the Thermodynamic Properties and the P, T Phase Diagram of Carbon," Carbon **24**, 169 (1986).

F. P. Bundy, "The P, T Phase and Reaction Diagram for Elemental Carbon, 1979," J. Geophys. Res. **85**, 6930 (1980).

F. P. Bundy, "An Update of the P, T Phase Diagram of Elemental Carbon," in *Solid State Physics under Pressure*, S. Minomura, ed. (KTK Scientific, Tokyo, 1985) p. 1.

J. E. Field, ed., *The Properties of Diamond* (Academic, London, 1979).

References

1. F. P. Bundy, J. Geophys. Res. **85,** 6930 (1980).
2. F. P. Bundy, in *Solid State Physics under Pressure*, S. Minomura, ed. (KTK Scientific, Tokyo, 1985) p. 1.
3. F. P. Bundy, Physica A **156**, 169 (1989).
4. J. Donohue, *The Structures of the Elements* (Wiley, New York, 1974) pp. 254–260.
5. A. G. Whittaker, Science **200**, 763 (1978).
6. R. B. Heimann, J. Kleiman, and N. M. Salansky, Nature **306**, 164 (1983).
7. F. P. Bundy, J. Chem. Phys. **38**, 631 (1963).
8. F. P. Bundy, J. Chem. Phys. **35**, 383 (1961).
9. C. S. Kennedy and G. C. Kennedy, J. Geophys. Res. **81**, 2467 (1976).
10. W. H. Gust, Phys. Rev. B **22**, 4744 (1980).
11. J. Kleiman, R. B. Heimann, D. Hawken, and N. M. Salansky, J. Appl. Phys. **56**, 1440 (1984).
12. L. F. Vereshchagin, E. N. Yakovlev, G. N. Stepanov, and B. V. Vinogradov, ZhETF Pis. Red. **16**, 382 (1972) [JETP Lett. **16**, 270 (1972)].
13. J. A. van Vechten, Phys. Rev. B **7**, 1479 (1973).
14. M. Pavlovskii, Fiz. Tverd. Tela **13**, 893 (1971) [Sov. Phys. Solid State **13**, 741 (1971)].
15. K. Kondo and T. J. Ahrens, Geophys. Res. Lett. **10**, 281 (1983).
16. M. T. Yin and M. L. Cohen, Phys. Rev. B **29**, 6996 (1984).
17. H. J. F. Jansen and A. J. Freeman, Phys. Rev. B **35**, 8207 (1987).
18. M. T. Yin and M. L. Cohen, Phys. Rev. Lett. **50**, 2006 (1983).

19. M. T. Yin, Phys. Rev. B **30**, 1773 (1984).
20. R. Biswas, R. M. Martin, R. J. Needs, and O. H. Nielsen, Phys. Rev. B **30**, 3210 (1984).
21. A. K. McMahan, Phys. Rev. B **30**, 5835 (1984).
22. R. Biswas, R. M. Martin, R. J. Needs, and O. H. Nielsen, Phys. Rev. B **35**, 9559 (1987).
23. S. Fahy and S. G. Louie, Phys. Rev. B **36**, 3373 (1987).
24. M. van Thiel and F. H. Ree, Int. J. Thermophys. **10**, 227 (1989).
25. D. A. Young and R. Grover, in *Shock Waves in Condensed Matter 1987*, S. C. Schmidt and N. C. Holmes, eds. (North-Holland, Amsterdam, 1988) p. 131.
26. M. A. Sheindlin, Teplofiz. Vys. Temp. **19**, 630 (1981) [High Temp. **19**, 467 (1981)].
27. M. A. Scheindlin, Mat. Res. Soc. Symp. Proc. **22**, 33 (1984).
28. A. Cezairliyan and A. P. Miiller, Bull. Am. Phys. Soc. **32**, 608 (1987).
29. F. P. Bundy, J. Chem. Phys. **38**, 618 (1963).
30. N. S. Fateeva and L. F. Vereshchagin, ZhETF Pis. Red. **13**, 157 (1971) [JETP Lett. **13**, 110 (1971)].
31. R. Grover, J. Chem. Phys. **71**, 3824 (1979).
32. J. S. Gold, W. A. Bassett, M. S. Weathers, and J. M. Bird, Science **225**, 921 (1984).
33. M. S. Weathers and W. A. Bassett, Phys. Chem. Minerals **15**, 105 (1987).
34. J. W. Shaner, J. M. Brown, C. A. Swenson, and R. G. McQueen, J. Phys. (Paris) **45**, C8-235 (1984).
35. M. Togaya, High Press. Res. **4**, 342 (1990).
36. G. Galli, R. M. Martin, R. Car, and M. Parrinello, Phys. Rev. Lett. **63**, 988 (1989).
37. H. R. Leider, O. H. Krikorian, and D. A. Young, Carbon **11**, 555 (1973).
38. G. Galli, R. M. Martin, R. Car, and M. Parrinello, Science **250**, 1547 (1990).
39. I. A. Korsunskaya, D. S. Kamenetskaya, and I. L. Aptekar´, Fiz. Met. Metalloved. **34**, 942 (1972) [Phys. Met. Metallog. **34**, 39 (1972)].
40. A. Ferraz and N. H. March, Phys. Chem. Liq. **8**, 289 (1979).
41. J. Z. Hu and I. L. Spain, Solid State Comm. **51**, 263 (1984).
42. J. Z. Hu, L. D. Merkle, C. S. Menoni, and I. L. Spain, Phys. Rev. B **34**, 4679 (1986).
43. F. P. Bundy, J. Chem. Phys. **41**, 3809 (1964).
44. T. I. Dyuzheva, S. S. Kabalkina, and V. P. Novichkov, Zh. Eksp. Teor. Fiz. **74**, 1784 (1978) [Sov. Phys. JETP **47**, 931 (1978)].
45. H. Olijnyk, S. K. Sikka, and W. P. Holzapfel, Phys. Lett. **103A**, 137 (1984).
46. S. J. Duclos, Y. K. Vohra, and A. L. Ruoff, Phys. Rev. B **41**, 12021 (1990).
47. V. Vijaykumar and S. K. Sikka, High Press. Res. **4**, 306 (1990).
48. S. J. Duclos, Y. K. Vohra, and A. L. Ruoff, Phys. Rev. Lett. **58**, 775 (1987).
49. J. S. Kasper and S. M. Richards, Acta Cryst. **17**, 752 (1964).
50. W. H. Gust and E. B. Royce, J. Appl. Phys. **42**, 1897 (1971).
51. A. Morita and T. Soma, Solid State Comm. **11**, 927 (1972).
52. K. J. Chang and M. L. Cohen, Phys. Rev. B **31**, 7819 (1985).
53. R. J. Needs and R. M. Martin, Phys. Rev. B **30**, 5390 (1984).
54. A. K. McMahan and J. A. Moriarty, Phys. Rev. B **27**, 3235 (1983).
55. F. Zandiehnadem and W. Y. Ching, Phys. Rev. B **41**, 12162 (1990).
56. R. Biswas and M. Kertesz, Phys. Rev. B **29**, 1791 (1984).
57. T. Soma and H. Matsuo, J. Phys. C **15**, 1873 (1982).

58. G. L. Warren and W. E. Evenson, Phys. Rev. B **11**, 2979 (1975).
59. I. Stich, R. Car, and M. Parrinello, Phys. Rev. Lett. **63**, 2240 (1989).
60. C. S. Menoni, J. Z. Hu, and I. L. Spain, Phys. Rev. B **34**, 362 (1986).
61. Y. K. Vohra, K. E. Brister, S. Desgreniers, A. L. Ruoff, K. J. Chang, and M. L. Cohen, Phys. Rev. Lett. **56**, 1944 (1986).
62. W. H. Gust and E. B. Royce, J. Appl. Phys. **43**, 4437 (1972).
63. K. J. Chang and M. L. Cohen, Phys. Rev. B **34**, 8581 (1986).
64. S. N. Vaidya, J. Akella, and G. C. Kennedy, J. Phys. Chem. Solids **30**, 1411 (1969).
65. I. N. Nikolaev, V. P. Mar´in, V. N. Panyushkin, and L. S. Pavlyukov, Fiz. Tverd. Tela **14**, 2337 (1972) [Sov. Phys. Solid State **14**, 2022 (1973)].
66. J. D. Barnett, V. E. Bean, and H. T. Hall, J. Appl. Phys. **37**, 875 (1966).
67. A. I. Kingon and J. B. Clark, High Temp.-High Press. **12**, 75 (1980).
68. H. Olijnyk and W. B. Holzapfel, J. Phys. (Paris) **45**, C8-153 (1984).
69. M. Liu and L. Liu, High Temp.-High Press. **18**, 79 (1986).
70. S. Desgreniers, Y. K. Vohra, and A. L. Ruoff, Phys. Rev. B **39**, 10359 (1989).
71. J. Ihm and M. L. Cohen, Phys. Rev. B **23**, 1576 (1981).
72. A. Jayaraman, W. Klement, Jr., and G. C. Kennedy, Phys. Rev. **130**, 540 (1963).
73. T. Takahashi, H. K. Mao, and W. A. Bassett, Science **165**, 1352 (1969).
74. H. Mii, I. Fujishiro, M. Senoo, and K. Ogawa, High Temp.-High Press. **5**, 155 (1973).
75. C. A. Vanderborgh, Y. K. Vohra, H. Xia, and A. L. Ruoff, Phys. Rev. B **41**, 7338 (1990).
76. N. E. Christensen, S. Satpathy, and Z. Pawlowska, Phys. Rev. B **34**, 5977 (1986).
77. P. W. Mirwald and G. C. Kennedy, J. Phys. Chem. Solids **37**, 795 (1976).
78. K. Inoue, A. Nakaue, and Y. Yagi, in *Proceedings of the 4th International Conference on High Pressure*, J. Osugi, ed. (Kawakita, Kyoto, 1975) p. 757.
79. B. K. Godwal, C. Meade, R. Jeanloz, A. Garcia, A. Y. Liu, and M. L. Cohen, Science **248**, 462 (1990).
80. H. Bernier and P. Lalle, in *High Pressure in Research and Industry*, C.-M. Backman, T. Johannisson, and L. Tegnér, eds. (Arkitektkopia, Uppsala, 1982) p. 194.
81. D. A. Boness, J. M. Brown, and J. W. Shaner, in Ref. 25, p. 115.
82. J. L. Pelissier, Physica **126A**, 271 (1984).
83. A. Y. Liu, K. J. Chang, and M. L. Cohen, Phys. Rev. B **37**, 6344 (1988).

CHAPTER 9
The Group V Elements

9.1 Introduction

The Group V elements do not have any unique technological importance, and they have received rather less attention than other groups. The trend toward insulating behavior continues in these elements with the appearance of N in diatomic molecular form. Also, metallic behavior increases either with increasing atomic weight or with increasing pressure, as usual.

9.2 Nitrogen

Nitrogen at low temperatures is a diatomic molecular solid or liquid. Below 35.6 K, α-N_2 has the Pa3 sc(4) structure, with the molecular axes pointing along the body diagonals of the cubic unit cell[1]. This is shown in Fig. 4.1. The molecules are orientationally ordered and undergo orientational oscillations (librations) as well as center-of-mass vibrations about their equilibrium positions. Between 35.6 K and the melting point at 63.2 K, N_2 has the β hcp structure. This structure shows rotational disorder[1]. From XRD data alone it cannot be determined whether the molecules are precessing or are librating in orientationally disordered states[2]. Neutron-diffraction data for single-crystal β-N_2 suggest nearly random orientational order[3]. The molecular ordering of β-N_2 is not sensitive to pressure[4].

At low temperature and 0.35 GPa, the α sc(4) phase transforms to the tetragonal γ st(2) phase. Here the molecules are arranged in layers with common orientation, each layer having a 90° shift in orientation from neighboring layers[2]. The α-β-γ triple point occurs at 44.5 K and 0.46 GPa[5]. In each of the α, β, and γ phases there are ambiguities in the XRD patterns which make the precise space group uncertain. However, it frequently happens that alternate structures which are crystallographically allowed are unlikely from a packing-energy standpoint.

The δ phase of N_2 was first discovered at 300 K and 4.9 GPa[6]. It has the Pm3n sc(8) crystal structure, which is isomorphic with the high-temperature forms of O_2 and F_2. In the unit cell, there are 6 molecules with disklike disorder, and 2 with spherical disorder. The β-δ phase boundary has been measured out to a triple point on the melting curve at ca. 8.9 GPa[7–10]. There are discrepancies among the experimental data, probably due to different ruby-fluorescence pressure scales.

By analogy with the γ-β O_2 and β-α F_2 transitions, it was expected that if δ-N_2 were cooled at a fixed high pressure, it would transform by distortion to a rotationally ordered rhombohedral lattice[11]. Raman data indicate a transition from γ-N_2 to a new phase (ε) at 1.9 GPa[8], and XRD measurements show this to be the rhombohedral $R\bar{3}c$, rh(8) structure[12]. The δ-ε transition occurs through the orientational ordering and displacement of the N_2 molecules. The δ-ε transition has been measured over the range 100–300 K by XRD[12]. A further transition at 15 K near 20 GPa has been observed by Raman spectroscopy. This phase, named ζ, is also a lower-symmetry distortion of the cubic δ phase, probably the rhombohedral R3c symmetry[8]. A Raman vibron splitting feature at RT above 16 GPa may correspond to another new phase, η[9,13–15]. RT DAC measurements using Brillouin and Raman spectroscopy have identified further anomalies at ca. 66 GPa and 100 GPa[14,15]. All of these structures are closely related to δ-N_2, as indicated by the smooth continuation of the δ Raman spectrum. In two experiments at RT, N_2 has been compressed to 130 and 180 GPa, and although it takes on a dark color at these pressures, the presence of the N_2 vibron lines shows that it is not yet dissociated[14,16].

N_2 is much less chemically reactive than O_2 and the halogens, and is therefore both experimentally and theoretically convenient for the study of diatomic phase diagrams. Since N_2 is a sufficiently massive molecule that its motion can be treated as nearly classical, the theoretical analysis is greatly facilitated. In diatomic molecular systems, the problem of calculating the phase diagram is made difficult both by the complexity of the intermolecular potential and by the complexity of the molecular motions. These problems have so far prevented the kind of quantitatively accurate predictions found in the rare gases and simple metals.

There have been numerous attempts to parameterize the N_2-N_2 potential[17,18]. These have been based on *ab initio* calculations as well as on fits to experimental data. As phase-transition pressures have increased, the fitted potentials have increasingly emphasized accurate short-range intermolecular repulsions. Representations of the potential typically assume either a single interaction center with explicit angular dependence or a two- (or multi-) center model in which the angle dependence is implicit. The latter models are referred to as atom-atom or site-site potentials. The electric multipole moments are included or not, depending on the object of the calculation.

The low-pressure portion of the N_2 phase diagram is closely analogous to that of $J = 1$ H_2, and the theoretical explanation is the same, namely strong electric quadrupolar interactions. Numerous theoretical calculations have been made of the static and dynamic properties of α- and β-N_2[17]. These models typically involve some form of lattice dynamics, which assumes small oscillations. However, the orientational oscillations in α-N_2 are large, suggesting the use of self-consistent phonon theory[19]. In β-N_2, MC and MD calculations reveal librations interrupted by large-angle reorientations[20–22]. This theoretical result disproves the precession model inferred from XRD[2], and also indicates that the lattice-dynamics formalism for librations is invalid. In addition, the potentials used in the calculations frequently omit the electric-quadrupole interaction for simplicity. Thus, theoretical attempts to model the complex motions in α- and β-N_2 and to calculate the phase transition between them have had some success, but the results are not yet completely convincing[19,23]. Perhaps the best result so far has been obtained from a Monte Carlo study of an electric-quadrupole lattice, which predicts the α-β transition close to the observed temperature[24].

The α-γ transition has also proven difficult to predict theoretically[17]. Atom-atom potentials tend to show α as more stable even at very high pressures. Only a potential with a Kihara-type spheroidal repulsive core has been fitted to the transition[25,26]. Even here, different quadrupole moments have to be used for the potentials in the two phases. This suggests that the simple atom-atom model of the repulsive potential is not sufficiently realistic. It is significant that there is no γ phase in carbon monoxide, which is isoelectronic with N_2[12].

The discovery of new high-pressure, low-temperature phases of N_2 has stimulated a major theoretical effort. Zero-Kelvin static-lattice calculations based on the Kim-Gordon electron-gas theory of intermolecular potentials have been used to find the lowest-energy crystal structures of N_2[27–29]. These calculations, using added attractive dispersion potential terms, correctly predict the γ-ε transition near 2.0 GPa[28,29]. Isobaric-isothermal MD calculations have been used to study the preferred structure taken by classical N_2 molecules interacting with atom-atom potentials by cooling the δ-N_2 phase along an isobar in steps[22,30,31]. Depending on the potential used, these calculations predict either the tetragonal st(64) or the rhombohedral R3c rh(64) structures at low temperature and 7.0 GPa. Although these predictions are incorrect, the calculations illustrate the utility of MD for "natural" searches for the most stable structures. A lattice-dynamics program[32] which carries out a search of crystal structures for the most stable phase has been conducted with various atom-atom potentials[11,33,34]. These calculations predicted the γ-ε transition at 1.92 GPa, in very good agreement with experiment[33]. The sensitivity of the predicted high-pressure phases to the assumed potential is a common feature of all of these calculations.

The insulator-metal transition in N_2 has been studied by separate calculations for the molecular $R\bar{3}c$ and metallic-phase total energies[35]. Using LMTO, which is restricted to the more symmetric structures, the most stable metallic phase is found to be sc(1). The static-lattice transition pressure is 70 GPa, with a very large (35%) volume change[35]. Phonon corrections could bring the transition pressure up to ca. 100 GPa. Extensive AIP total-energy calculations on the high-pressure dissociation of N_2 find an open metallic structure, A7 rh(2), and a transition pressure of ca. 100 GPa[36]. Both calculations predict a first-order insulator-metal transition, unlike that predicted for H_2. There is at present no theoretical explanation for the continued existence of the molecular phase up to 180 GPa.

In connection with the insulator-metal transition in nitrogen, it is interesting that shocked N_2 shows a principal Hugoniot with a marked "softening" above 30 GPa[37]. The softening leads to such anomalies as "shock cooling," negative values of $\partial P/\partial T$, and crossing isotherms. This is interpreted as dissociation into a dense atomic phase[38]. A chemical-equilibrium theory of the dissociation suggests that it may be a phase transition of first order, which joins smoothly with the solid-dissociation transition somewhere above 100 GPa[39].

The melting curve of N_2 has been measured to 17 GPa by optical and Raman measurements in a DAC especially designed for high temperatures. There is one triple point, β-δ-liquid, on this curve[9,10,40].

Because the molecular rotational motions in solid and liquid N_2 are different, there is a rotational free-energy difference which contributes to the pressure of the melting transition. This difference has the effect of raising the transition pressure above that predicted by a sphericalized average potential[40]. MC calculations on N_2 using a new "lattice-coupling expansion" method have obtained absolute free energies for the solid and liquid phases at 300 K[41]. The predicted melting volumes and pressure are in very good agreement with experiment. The N_2 phase diagram is shown in Fig. 9.1.

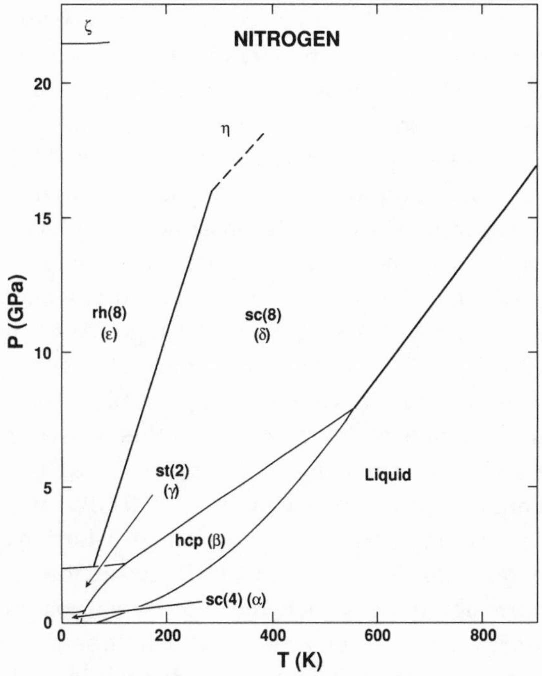

Fig. 9.1 The phase diagram of nitrogen.

N_2 has been the most thoroughly studied of the diatomic solids. The nonspherical molecular shape and internal degrees of freedom produce a phase diagram far more complex than those of the monatomic insulators such as Ar. The standard statistical-mechanical methods for calculation of thermodynamic functions have generally failed for N_2, while the MC and MD simulation methods have been more successful. Calculation of the diatomic phase diagrams remains a formidable theoretical challenge.

9.3 Phosphorus

At RTP P displays a number of allotropes, ranging from the "white" phase composed of P_4 molecules to the polymeric "black" phase (I), with orthorhombic structure eco(8)[1]. Thermodynamic measurements show that the black phase is the most stable of the various forms. Black P is unique in that it requires synthesis at high pressure and temperature. It is composed of puckered layers of covalently bonded atoms with weak van der Waals interlayer forces. It is a narrow-gap semiconductor, and compression to 1.7 GPa at RT closes the gap and gives metallic conduction without any change of crystal structure[42]. In its color, crystal structure, anisotropy, and melting behavior, black P is remarkably similar to graphitic carbon.

Compression of black P at RT and 5.5 GPa shows a transition to a semimetallic A7 rh(2) structure (II), which is the common low-pressure form of the heavier Group V elements[43]. Extensive high-temperature experiments have been performed with a cubic multianvil device. The I-II transition has a negative dP/dT and the phase boundary has been measured out to a triple point on the melting curve[44]. Further compression to 10 GPa at RT produces the metallic sc(1) III phase[43]. At low temperatures this transition is first-order with a 3.7% volume change. As the temperature is increased, the XRD reflections show a continuous transition from rh(2) to sc(1), which suggests that the volume change approaches zero above ca. 650 K. For this transition, $dP/dT \cong 0$[44]. No other phases are found up to 1400 K. The sc(1) phase remains stable up to 70 GPa[45]. There are indications of another phase change in the 80–100 GPa range[45].

Theoretical study of P has focused on the lattice distortions which lead to the two phase transitions[46,47]. The black P phase shows a strong anisotropy in compressibility, in which the interlayer separation decreases strongly with pressure, while the covalent bond length hardly changes[43,48]. The relationship among the I, II, and III structures is indicated in Fig. 9.2.

For the rh(2) and sc(1) structures, NFE pseudopotential theory has successfully predicted the first-order transition between them and the great stability range of the sc(1)[49,50]. These open structures are stabilized by the band-structure term in the total energy, which outweighs the Madelung term favoring the more closely packed lattices. NFE models not surprisingly fail to predict the stability of the eco(8) phase. AIP calculations clearly show the stability of the eco(8) phase at RP, and the transition to the rh(2) phase under pressure[47]. These calculations do not predict the transition to sc(1), however, and the authors have found that a constant shift in the rh(2) energy is needed to bring the theoretical pressures of the two transitions into agreement with experiment. More recent AIP calculations on the rh(2)-sc(1) transition obtain a transition pressure of 15.8 GPa without any correction[51]. This model also predicts a transition to bcc at 135 GPa.

At $P = 0$, black P melts to a molecular P liquid which does not freeze back to black P on experimental time scales. Equilibrium freezing occurs only at higher pressures[42]. The melting curve of P has been measured to 5.0 GPa in the multianvil cell[52]. XRD was used to record the transition. A melting-temperature maximum is clearly seen on the black P melting curve, indicating a shift in the liquid from a polyatomic insulating to a monatomic metallic structure. The phase diagram of P is shown in Fig. 9.3.

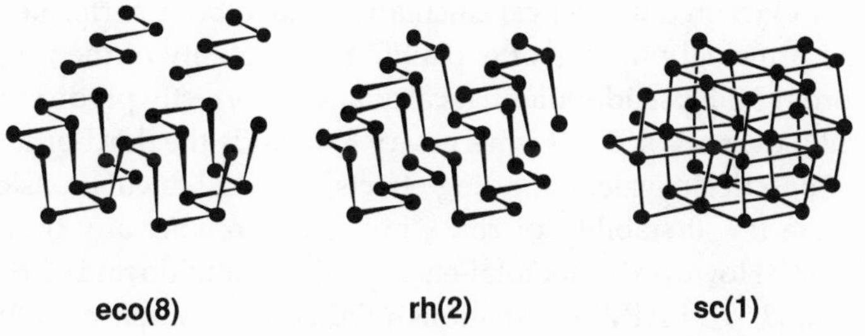

Fig. 9.2 Phosphorus crystal structures. (From Kilegawa and Iwasaki[43]. Redrawn with permission.)

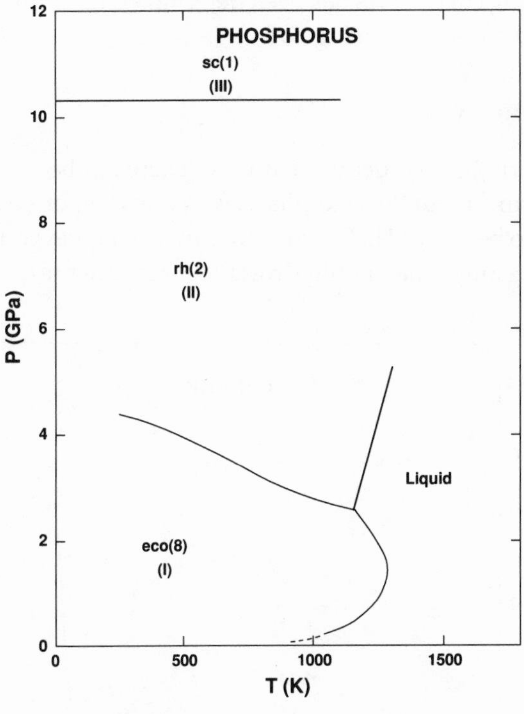

Fig. 9.3 The phase diagram of phosphorus.

9.4 Arsenic

At RTP As is a semimetal in the A7 rh(2) phase, isostructural with P II. XRD and Raman measurements in a DAC at RT show a transition from rh(2) to sc(1) at 25 GPa[53,54]. The transition is probably first-order, but the volume change is very small. No further transitions are found up to 45 GPa[53].

Rather extensive theoretical calculations have been performed for As, with the principal purpose of explaining the stability of the rh(2) phase. Third-order NFE pseudopotential calculations correctly predict the stability of rh(2) over sc(1)[55]. It is necessary to include the third-order term in order to describe covalent bonding effects[52]. AIP calculations correctly predict the $P = 0$ stability of rh(2), but fail to predict any transition to sc(1)[56,57]. However, other total-energy calculations do find an rh(2)-sc(1) transition[58,59]. LAPW calculations predict a transition pressure of ca. 19 GPa[58], and independent AIP calculations of the phonons in the sc(1) lattice suggest a transition volume ($V/V_0 = 0.72$) close to that observed (0.74)[59].

The melting curve of As has been determined to 6.0 GPa[60]. The phase diagram of As is shown in Fig. 9.4.

9.5 Antimony

At RP, Sb is rh(2), isostructural with As. There has been a lengthy dispute in the literature about the next phase, which was reported at 5.0 GPa at RT, and found to be sc(1)[61]. Recent experiments, however, have shown that sc(1) appears only when nonhydrostatic forces are present in the pressure

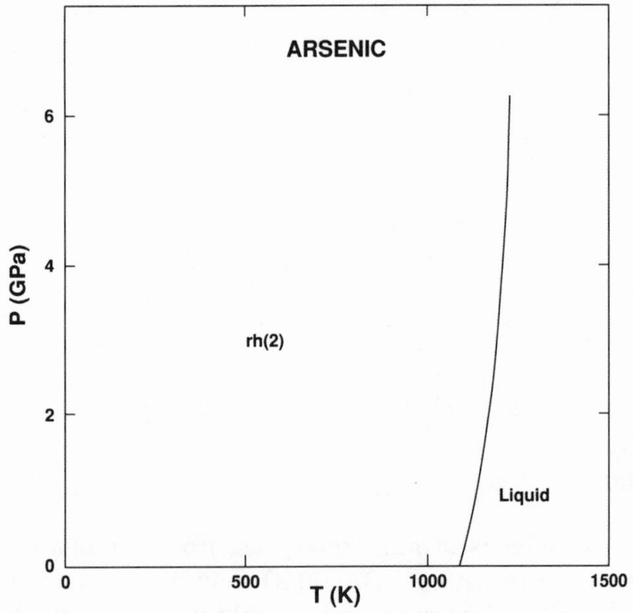

Fig. 9.4 The phase diagram of arsenic.

cell[62,63]. Although the rh(2) phase approaches the sc(1) structure, it never reaches it. The next phase, Sb II, occurs at 8.0 GPa. The structure of Sb II has been indexed as monoclinic with 8 atoms per unit cell[64]. The exact space group is undetermined. The I-II phase line has been followed to a triple point on the melting curve[64–66]. There is no sign of an sc(1) phase at any temperature[66]. Further phases (IV and V) have been claimed above 8.0 GPa, on the basis of superconductivity measurements[67] and by analogy with Bi, but the most recent XRD data show that Sb II persists to 28 GPa, where it transforms to bcc, phase III[68]. No further phases are reported to 43 GPa.

There has been comparatively little theoretical work on Sb. AIP calculations have shown that the sc(1) phase has higher energy than bcc above 28 GPa, as observed[69].

The melting curve has been measured to 8.0 GPa[60,65,66]. There is some disagreement among the measurements, but all show a clear cusp at the I-II-liquid triple point. The phase diagram of Sb is shown in Fig. 9.5.

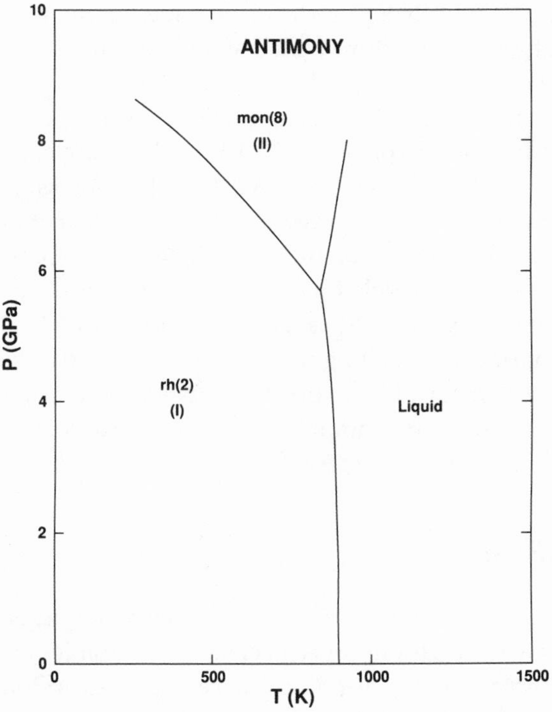

Fig. 9.5 The phase diagram of antimony.

9.6 Bismuth

Bi follows the general rule that the heavier elements in a series have a given transition at a lower pressure. However, Bi differs from the other Group V elements in having a much more complex phase diagram, with as many as 10 different phases. The literature on the Bi diagram is also confused by controversy over the existence of certain phases and by constant changes in phase numbering due to conflicting observations. Here the numbering due to Homan[70] is followed.

At RP, Bi I is rh(2), isomorphic with As and Sb. Compression at RT to 40 GPa produces the structure sequence I→II→III→IV→ V→VI[70]. Phase II has been identified as cm(4)[71]. Neutron-scattering measurements on Bi III have tentatively identified it as orthorhombic with 8 atoms per unit cell[72]. The high-pressure phase VI is bcc[73]. No further transitions are found up to 40 GPa[74].

At temperatures below RT, the I-II transition disappears in a triple point[75], and two new phases, VIII and IX, appear[70]. At high temperatures, two additional phases, VII and II′, with a very small stability field[76], are found. Neutron-diffraction measurements on phase VII indicate a tetragonal structure with 8 atoms per unit cell, very similar to phase III[72].

A theoretical calculation[74] using a NFE model has been carried out for the sc(1), bcc, fcc, and hcp phases of Bi. The calculation clearly shows a low-pressure sc(1) stability range followed by a high-pressure bcc stability range. The transition pressure predicted is roughly that of the observed V-VI transition. In the calculation the loosely packed low-pressure phases of Bi are approximately modeled with the sc(1) structure.

The melting curve of Bi has been measured to 7.0 GPa[62]. It is interesting that the rh(2) melting curve has a large negative slope, unlike the other Group V elements. Liquid Bi also exhibits a resistance anomaly which suggests a structural transition in the liquid ordering[77]. The phase diagram of Bi is shown in Fig. 9.6.

9.7 Discussion

Like the Group IV elements, the Group V elements show a general sequence of structures with variations for each element: diatomic→covalent→rh(2) or sc(1)→complex→bcc. The RT sequence of phases is shown in Fig. 9.7. The s^2p^3 electron configuration in the Group V elements promotes covalent bonding by the *p* electrons. This favors bond angles close to 90°, as seen in

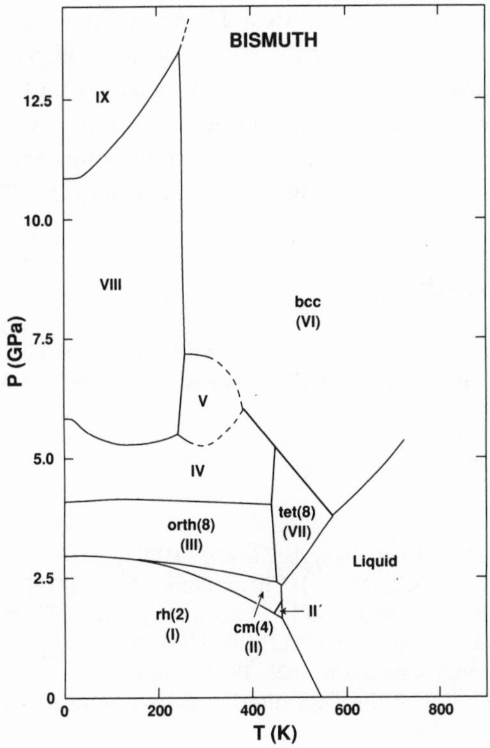

Fig. 9.6 The phase diagram of bismuth.

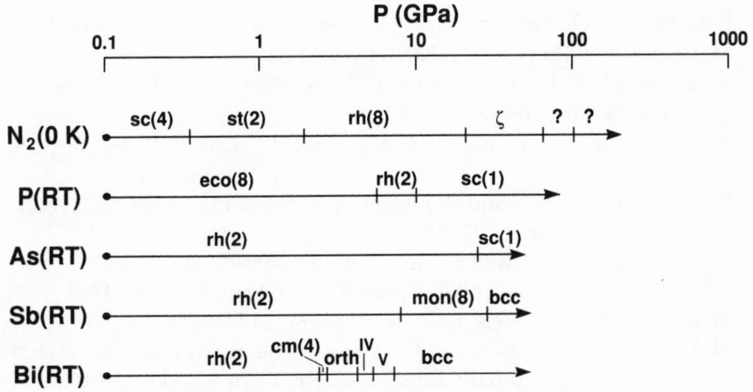

Fig. 9.7 Group V phase transitions.

the eco(8), rh(2), and sc(1) structures. The preference shown for rh(2) over sc(1) at low pressures is due to a Peierls distortion, which lowers the energy of the sc(1) lattice by splitting the electron bands and shifting electron density to lower energies[59]. The liquid state of the heavier Group V elements evidently retains the covalent semiconducting character of the solid phases, and this is eliminated only by high temperatures[77].

For Further Reading

T. A. Scott, "Solid and Liquid Nitrogen," Phys. Repts. **27**, 89 (1976).
A. Morita, "Semiconducting Black Phosphorus," Appl. Phys. A **39**, 227 (1986).

References

1. J. Donohue, *The Structures of the Elements* (Wiley, New York, 1974) chap. 8.
2. A. F. Schuch and R. L. Mills, J. Chem. Phys. **52**, 6000 (1970).
3. B. M. Powell, G. Dolling, and H. F. Nieman, J. Chem. Phys. **79**, 982 (1983).
4. D. Schiferl, D. T. Cromer, and R. L. Mills, High Temp.-High Press. **10**, 493 (1978).
5. C. A. Swenson, J. Chem. Phys. **23**, 1963 (1955).
6. D. T. Cromer, R. L. Mills, D. Schiferl, and L. A. Schwalbe, Acta Cryst. **B37**, 8 (1981).
7. S. Buchsbaum, R. L. Mills, and D. Schiferl, J. Phys. Chem. **88**, 2522 (1984).
8. D. Schiferl, S. Buchsbaum, and R. L. Mills, J. Phys. Chem. **89**, 2324 (1985).
9. A. S. Zinn, D. Schiferl, and M. F. Nicol, J. Chem. Phys. **87**, 1267 (1987).
10. W. L. Vos and J. A. Schouten, J. Chem. Phys. **91**, 6302 (1989).
11. K. Kobashi, A. A. Helmy, R. D. Etters, and I. L. Spain, Phys. Rev. B **26**, 5996 (1982).
12. R. L. Mills, B. Olinger, and D. T. Cromer, J. Chem. Phys. **84**, 2837 (1986).
13. M. Grimsditch, Phys. Rev. B **32**, 514 (1985).
14. R. Reichlin, D. Schiferl, S. Martin, C. Vanderborgh, and R. L. Mills, Phys. Rev. Lett. **55**, 1464 (1985).
15. A. P. Jephcoat, R. J. Hemley, H. K. Mao, and D. E. Cox, Bull. Am. Phys. Soc. **33**, 522 (1988).
16. P. M. Bell, H. K. Mao, and R. J. Hemley, Physica **139&140B**, 16 (1986).
17. T. A. Scott, Phys. Repts. **27**, 89 (1976).
18. M. S. H. Ling and M. Rigby, Mol. Phys. **51**, 855 (1984).
19. J. C. Raich, N. S. Gillis, and T. R. Koehler, J. Chem. Phys. **61**, 1411 (1974).
20. M. L. Klein and J.-J. Weis, J. Chem. Phys. **67**, 217 (1977).
21. M. L. Klein, D. Levesque, and J.-J. Weis, J. Chem. Phys. **74**, 2566 (1981).
22. J. Belak, R. LeSar, and R. D. Etters, J. Chem. Phys. **92**, 5430 (1990).
23. A. van der Avoird, W. J. Briels, and A. P. J. Jansen, J. Chem. Phys. **81**, 3658 (1984).
24. M. J. Mandell, J. Chem. Phys. **60**, 4880 (1974).
25. J. C. Raich and R. L. Mills, J. Chem. Phys. **55**, 1811 (1971).
26. K. Kobashi and T. Kihara, J. Chem. Phys. **72**, 378 (1980).

27. R. LeSar and R. G. Gordon, J. Chem. Phys. **78**, 4991 (1983).
28. R. LeSar, J. Chem. Phys. **81**, 5104 (1984).
29. R. LeSar and M. S. Shaw, J. Chem. Phys. **84**, 5479 (1986).
30. S. Nosé and M. L. Klein, Phys. Rev. Lett. **50**, 1207 (1983).
31. S. Nosé and M. L. Klein, Phys. Rev. B **33**, 339 (1986).
32. R. Pan and R. D. Etters, J. Chem. Phys. **72**, 1741 (1980).
33. V. Chandrasekharan, R. D. Etters, and K. Kobashi, Phys. Rev. B **28**, 1095 (1983).
34. R. D. Etters, V. Chandrasekharan, E. Uzan, and K. Kobashi, Phys. Rev. B **33**, 8615 (1986).
35. A. K. McMahan and R. LeSar, Phys. Rev. Lett. **54**, 1929 (1985).
36. R. M. Martin and R. J. Needs, Phys. Rev. B **34**, 5082 (1986).
37. H. B. Radousky, W. J. Nellis, M. Ross, D. C. Hamilton, and A. C. Mitchell, Phys. Rev. Lett. **57**, 2419 (1986).
38. M. Ross, J. Chem. Phys. **86**, 7110 (1987).
39. D. C. Hamilton and F. H. Ree, J. Chem. Phys. **90**, 4972 (1989).
40. D. A. Young, C.-S. Zha, R. Boehler, J. Yen, M. Nicol, A. S. Zinn, D. Schiferl, S. Kinkead, R. C. Hanson, and D. A. Pinnick, Phys. Rev. B **35**, 5353 (1987).
41. E. J. Meijer, D. Frenkel, R. A. LeSar, and A. J. C. Ladd, J. Chem. Phys. **90**, 7570 (1990).
42. A. Morita, Appl. Phys. A **39**, 227 (1986).
43. T. Kikegawa and H. Iwasaki, Acta Cryst. **B39**, 158 (1983).
44. T. Kikegawa, H. Iwasaki, T. Fujimura, S. Endo, Y. Akahama, T. Akai, O. Shimomura, T. Yagi, S. Akimoto, and I. Shirotani, J. Appl. Cryst. **20**, 406 (1987).
45. M. Okajima, S. Endo, Y. Akahama, and S. Narita, Jpn. J. Appl. Phys. **23**, 15 (1984).
46. J. K. Burdette and S. L. Price, Phys. Rev. B **25**, 5778 (1982).
47. K. J. Chang and M. L. Cohen, Phys. Rev. B **33**, 6177 (1986).
48. L. Cartz, S. R. Srinavasa, R. J. Riedner, J. D. Jorgensen, and T. G. Worlton, J. Chem. Phys. **71**, 1718 (1979).
49. D. Schiferl, Phys. Rev. B **19**, 806 (1979).
50. A. Morita and H. Asahina, in *Solid State Physics under Pressure*, S. Minomura, ed. (KTK Scientific, Tokyo, 1985) p. 197.
51. T. Sasaki, K. Shindo, K. Niizeki, and A. Morita, J. Phys. Soc. Jpn. **57**, 978 (1988).
52. Y. Akahama, W. Utsumi, S. Endo, T. Kikegawa, H. Iwasaki, O. Shimomura, T. Yagi, and S. Akimoto, Phys. Lett. A **122**, 129 (1987).
53. T. Kikegawa, and H. Iwasaki, J. Phys. Soc. Jpn. **56**, 3417 (1987).
54. H. J. Beister, K. Strössner, and K. Syassen, Phys. Rev. B **41**, 5535 (1990).
55. A. Morita, I. Ohkoshi, and Y. Abe, J. Phys. Soc. Jpn. **43**, 1610 (1977).
56. R. J. Needs, R. M. Martin, O. H. Nielsen, Phys. Rev. B **33**, 3778 (1986).
57. R. J. Needs, R. M. Martin, and O. H. Nielsen, Phys. Rev. B **35**, 9851 (1987).
58. L. F. Mattheiss, D. R. Hamann, and W. Weber, Phys. Rev. B **34**, 2190 (1986).
59. K. J. Chang and M. L. Cohen, Phys. Rev. B **33**, 7371 (1986).
60. W. Klement, Jr., A. Jayaraman, and G. C. Kennedy, Phys. Rev. **131**, 632 (1963).
61. S. S. Kabalkina and V. P. Mylov, Doklady Akad. Nauk SSSR **152**, 585 (1963) [Sov. Phys.-Doklady **8**, 917 (1964)].
62. L. G. Khvostantsev and V. A. Sidorov, Phys. Stat. Sol. (a) **64**, 379 (1981).
63. D. Schiferl, D. T. Cromer, and J. C. Jamieson, Acta Cryst. **B37**, 807 (1981).
64. H. Iwasaki and T. Kikegawa, Physica **139&140B**, 259 (1986).

65. S. A. Ivakhnenko and Ye. G. Ponyatovskiy, Fiz. Metall. Metalloved. **47**, 1314 (1979) [Phys. Met. Metall. **47**, 172 (1980)].
66. L. G. Khvostantsev and V. A. Sidorov, Phys. Stat. Sol. (a) **82**, 389 (1984).
67. M. A. Il´ina, Fiz. Tverd. Tela **22**, 849 (1980) [Sov. Phys. Solid State **22**, 494 (1980)].
68. K. Aoki, S. Fujiwara, and M. Kusakabe, Solid State Comm. **45**, 161 (1983).
69. T. Sasaki, K. Shindo, and K. Niizeki, Solid State Comm. **67**, 569 (1988).
70. C. G. Homan, J. Phys. Chem. Solids **36**, 1249 (1975).
71. R. M. Brugger, R. B. Bennion, and T. G. Worlton, Phys. Lett. **24A**, 714 (1967).
72. V. K. Fedotov, E. G. Ponyatovskii, V. A. Somenkov, and S. Sh. Shil´shtein, Fiz. Tverd. Tela **20**, 1088 (1978) [Sov. Phys. Solid State **20**, 628 (1978)].
73. Ph. Schaufelberger, H. Merx, and M. Contré, High Temp.-High Press. **5**, 221 (1973).
74. K. Aoki, S. Fujiwara, and M. Kusakabe, J. Phys. Soc. Jpn. **51**, 3826 (1982).
75. E. M. Compy, J. Appl. Phys. **41**, 2014 (1970).
76. N. A. Tikhomirova, E. Yu. Tonkov, and S. M. Stishov, ZhETF Pis´ma **3**, 96 (1966) [JETP Lett. **3**, 60 (1966)].
77. G. C. Vezzoli, High Temp.-High Press. **12**, 195 (1980).

CHAPTER 10
The Group VI Elements

10.1 Introduction

In the Group VI elements, the trend toward covalency is carried nearly to completion. All of these elements are covalent insulators or semiconductors except for Po, which is metallic. As usual, the molecular solids, especially O_2 and S, show complex phase diagrams. All of the Group VI elements have been metallized by application of pressure. The metallic structures, so far as they have been determined, tend to be open, as in Group V.

10.2 Oxygen

Solid diatomic oxygen exists in 3 phases at RP. Because the electronic ground state of O_2 is a triplet with spin $S = 1$, solid O_2 is the only magnetic insulator among the elements. Below 23.7 K, α-O_2 has a monoclinic structure cm(2) with 2 molecules per unit cell[1]. The monoclinic angle is 132°, and the O_2 molecules are oriented perpendicular to the close-packed layer planes. The α-O_2 phase is antiferromagnetic, with two sublattices in the layer planes[2]. The α-O_2 structure is shown in Fig. 10.1.

For $23.7 \leq T \leq 43.8$ K, β-O_2 has a rhombohedral structure rh(1) with 1 molecule in the unit cell[1]. The structure is only a slightly distorted form of α-O_2. Unlike α-O_2, β-O_2 has no long-range magnetic order. Various short-range orderings have been suggested[2-4], but the best fit to neutron scattering is a two-dimensional helicoidal order[4].

For $43.8 \leq T \leq 54.4$ K, γ-O_2 is a paramagnetic sc(8) structure with two crystallographically distinct molecular positions[1]. This structure is equivalent to δ-N_2. γ-O_2 is completely unlike α and β and the β-γ transition shows a very large volume change[5].

The high-pressure phase diagram has been explored only recently. At low temperatures, the molecular vibron frequency shows a distinct change

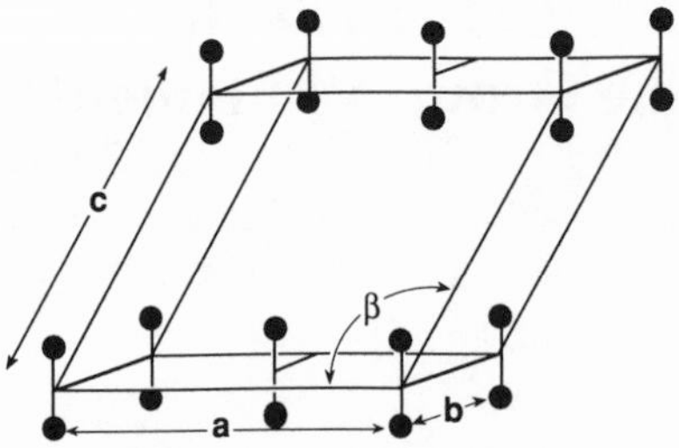

Fig. 10.1 The crystal structure of α-O_2. (From Etters, et al.[16]. Redrawn with permission.)

of slope at about 0.7 GPa[6,7]. Whether this is a first-order phase transition (to δ′-O_2) is not yet clear. At ca. 3 GPa and low temperature there is a definite phase change to an orthorhombic fco(4) structure[8,9], δ-O_2. This structure is closely related to that of α-O_2. It has an orange color in transmitted light. Further compression to ca. 8 GPa leads to the red ε phase. There has been much discussion on the structure of ε-O_2, but the difficulty of growing good crystals has prevented accurate XRD work until recently. The ε structure has been indexed to the same space group as α-O_2, namely C2/m or cm(2), with monoclinic angle 116.1°[10]. The principal difference between the α and ε phases is the much smaller intermolecular distance within the ε layer planes. This leads to overlap of unfilled orbitals with the resulting optical anisotropy and strong optical absorption[11]. Compression into the 100 GPa range changes the color of ε-O_2 to black, but reveals no change of phase[11,12]. Optical-absorption and reflectivity measurements indicate a band-gap closure and metallization at ca. 95 GPa[12].

The phase boundaries joining these solid phases have been worked out by means of Raman spectroscopy and XRD[6,7,9,13]. Two new, presumably metastable, phases, χ and ω, have been found at RT[9].

The theoretical analysis of the O_2 phase diagram has focused on the magnetic properties at RP and the crystal structures under compression at $T = 0$ K. It has been recognized for many years that the α-β phase change is a magnetoelastic phenomenon[14]. That is, in order to stabilize the two-sublattice antiferromagnetic state, the crystal lattice must distort to the lower monoclinic symmetry. If there were no β phase, the α phase would show a Néel point at ca. 31 K and become paramagnetic[3]. Similarly, if there were no α phase, the short-range antiferromagnetic phase would switch over to a long-range three-sublattice phase at ca. 21 K[3]. Theoretical calculations with spin-dependent terms in the Hamiltonian have convincingly predicted the observed behavior of α- and β-O_2[3,15–18]. Two recent theoretical predictions of the α-β transition temperature have been made. A lattice-dynamics model predicts 45 K[19] while a constant-pressure MD calculation predicts 17.8 K[18]. The latter calculation clearly exhibits the first-order character of the transition. The γ-O_2 phase has not received the same attention as α- and β-O_2, but a MD simulation[20] of this phase has shown it to exhibit rapid reorientational motions, similar to those found in the liquid state.

The sequence of phases found in O_2 at 0 K has been studied with a model which searches for the lowest-energy configuration of the lattice at any given volume[21]. The energy is computed as a static-lattice sum over pair interactions consisting of repulsive cores, van der Waals attractions, electric multipoles, and magnetic dipoles. Once an optimum configuration has been found, a lattice-dynamics calculation is run to find the zero-point energy and to carry out a local configurational search for the absolute energy minimum. These calculations are sensitively dependent on the details of the model potential. Even so, the agreement with experiment on the α-δ' and α-δ transition pressures is very good[21].

The melting curve of O_2 has been determined to 16.5 GPa[13,22]. Two triple points, the β-γ-liquid and β-ε-liquid, have been found. The O_2 melting curve is remarkable in its very large dP/dT. Near RT, the curve actually crosses that of H_2. This feature has been qualitatively explained as the result of strongly hindered rotation in the solid, which may be the result of intermolecular charge-transfer bonding[22]. The phase diagram of O_2 is shown in Fig. 10.2.

10.3 Sulfur

Sulfur has the most complex phase diagram of all the elements, and is one of the most difficult elements to investigate because of very long phase-equilibrium times. In addition, there are numerous metastable phases based on the molecular species S_6, S_7, S_8, S_9, S_{10}, S_{12}, S_{18}, and S_∞[1,23].

At RP the stable molecular species is S_8, which forms a slightly asymmetric puckered ring in its crystal environment. At temperatures below 360 K, the stable structure is orthorhombic (I or α) fco(16) with 16 S_8 molecules per unit cell. The packing is complex. Between 360 K and melting at 392 K, the stable structure is monoclinic (II or β) sm(6), with 6 S_8 molecules per unit cell. Four of the six molecules are orientationally ordered and two are disordered. These crystallographically distinct types of molecules have slightly different bond lengths and angles[1]. The S_8 ring in S I is shown in Fig. 10.3.

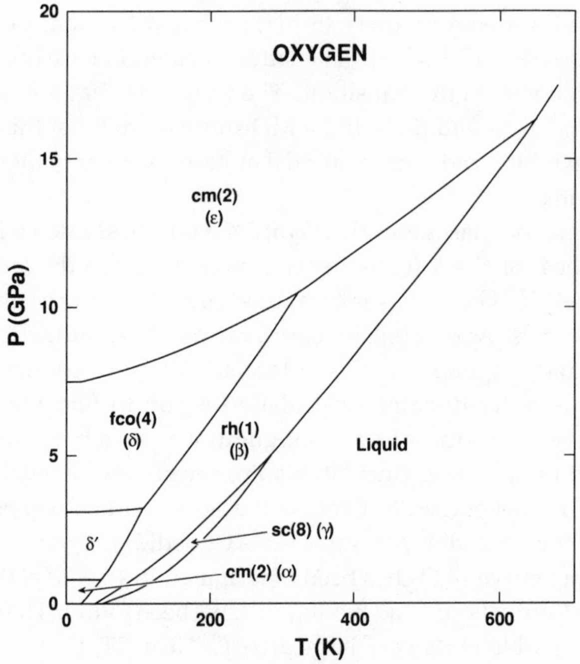

Fig. 10.2 The phase diagram of oxygen.

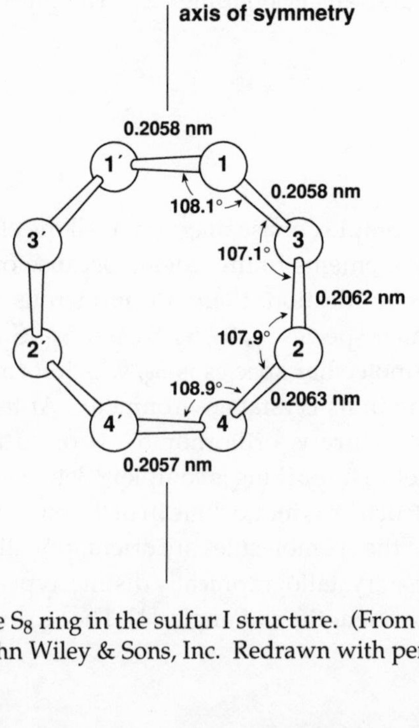

Fig. 10.3 The S_8 ring in the sulfur I structure. (From Donohue[1].

Given the variable and complex shape of the S_8 rings, it is evident that numerous phases might be possible as the pressure is increased and the molecules are forced closer together. In fact, 10 new solid phases, III-XII, have been reported below 5.0 GPa. Lengthy investigations which have made use of DTA, electrical conductivity measurements, and XRD study of samples recovered by rapid quenching, have outlined the phase boundaries in S[24].

Other than the RP phases I and II, only phase XII has been characterized structurally[24,25]. The XRD evidence for phase XII suggests a monoclinic cell with close-packed helical chains, similar to those in Se and Te. A nearly constant-temperature phase boundary found near 400 K may represent a ring-to-helix transition, with the $T < 400$ K phases having S_8 rings and the $T > 400$ K phases having helical S chains[26].

Two sets of DAC Raman spectroscopy experiments have yielded conflicting results. In one experiment, no phase transitions are found up to 5 GPa, where most of the Raman peaks vanish[27]. Single-crystal XRD measurements by the same authors suggest a transition from the fco(16) structure to a triclinic structure at 5 GPa. An independent Raman study to 50 GPa finds evidence of several phase transitions above 10 GPa[28]. Visual observation of the sample in the 10 GPa pressure range finds a color change from yellow to black, implying a major change in band structure. There is evidence that phase changes in S may be driven by laser light[29]. Obviously, continued experimental work is needed to clarify the S phase diagram.

At the highest pressures it looks as though phase XII is "taking over" the phase diagram[26]. At still higher pressures, there is a sharp drop in the resistivity of S, and it becomes metallic. There has been controversy about the transition pressure[30,31], but the most recent DAC measurements show the band gap decreasing to zero at 48 GPa[32]. Resistance measurements at high pressures indicate metallic behavior at high temperatures and semiconducting behavior at low temperatures[31,33]. A sharp drop in resistance at $P > 50$ GPa and $T = 870$ K may indicate melting[31]. The structure of the high-pressure metallic phase has not yet been determined.

Several shock-wave studies on S show a decrease in resistance and metallic behavior above 20 GPa[34–36]. This state is probably in the liquid, which means that liquid S passes through the insulator-to-metal transition at a lower pressure than in the solid.

Theoretical work on S has concentrated on the comparison of lattice-dynamics calculations with experimental vibrational spectra and thermodynamic data for S I at low pressure[37–39]. The comparisons show that the atom-atom model for the interaction of the S_8 rings is satisfactory. There have not yet been any attempts to predict phase transitions in solid S.

The melting curve of S has been determined to about 3.5 GPa[40]. Three cusps corresponding to solid-solid-liquid triple points have been found. A very unusual aspect of the S phase diagram is the evidence from DTA measurements that there are different liquid states or "phases" separated by narrow transition regions[41]. Only the RP transition at 433 K has been studied closely, and this is identified as a ring-chain polymerization. The transition is characterized by a lambda singularity in the specific heat[42] and is accurately modeled with a nearest-neighbor Ising lattice[43]. At ca. 460 K, there is another "phase" boundary representing the onset of depolymerization. Altogether 5 "phases" have been discovered. Other than phase A, these are evidently different polymeric states differing in chain length and spatial order. Thus the complexity of the solid S phase diagram extends into the liquid. The phase diagram of S is shown in Fig. 10.4.

10.4 Selenium

At RTP, Se I is rh(3). The atoms are arranged in infinite helices parallel to the c axis, with 3 atoms per turn[1], as shown in Fig. 10.5. This form is a semiconductor. There are also metastable monoclinic forms of Se based on Se_8 rings, similar to S_8[1]. Prior to recent XRD work on Se at high pressure, there was much disagreement about the structures of the high-pressure phases and their electrical conductivity[44–46]. DAC XRD measurements show three transitions, rh(3)→mon(3) (Se II) at 14 GPa; mon(3)→tet(4) (Se III) at 28 GPa; and tet(4)→rh(2) (Se IV) at 41 GPa[47]. The precise space groups are not reported. No further transitions are seen up to 50 GPa[47]. Although the I-II transition has a large (20%) volume change, it appears that the II-III transition is the insulator-metal transition[46].

Theoretical work on Se has focused mainly on comparison with experimental optical spectra. However, AIP structural calculations have been performed[48] which demonstrate the stability of the Se chain configuration. AIP calculations also show that the interchain interactions are strong and that normal band-structure analysis is valid[49]. There has been no published theoretical work yet on the high-pressure phases of Se.

The melting curve of Se has been determined to 6.0 GPa, and it shows a clear temperature maximum[50–52]. A more recent study shows a maximum nearer 10 GPa[53]. The behavior of liquid Se, like that of S, is complex. Near the melting curve at RP, NMR studies show the liquid to be a semiconductor composed of long chains, as in the solid[54]. As the

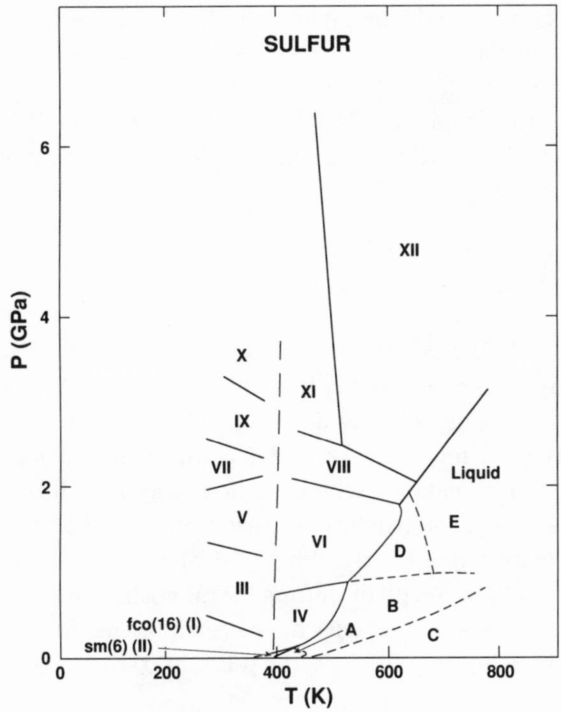

Fig. 10.4 The phase diagram of sulfur.

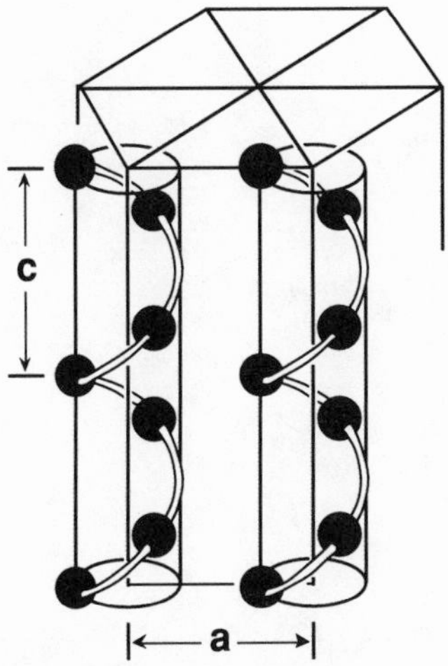

Fig. 10.5 The crystal structure of selenium I. (From H. M. Isomäki and J. von Boehm, Phys. Rev. B **35**, 8019 [1987]. Redrawn with permission.)

temperature is increased, the chains shorten, and the fluid becomes a paramagnetic insulator. At high pressures, the coordination number in the liquid increases smoothly[53]. There is evidence for a sharp metal-non-metal transition in the liquid at 3.6 GPa on the melting curve[55]. The phase boundary for this transition has a negative dP/dT. The Se phase diagram is shown in Fig. 10.6.

10.5 Tellurium

At RTP, Te(I) is rh(3), a semiconductor isomorphic with Se. However, the lattice c/a ratio differs substantially from that of Se, so that the Te lattice may be described as an only slightly distorted sc(1) structure[1]. A series of structures has been found in Te under compression at RT[56–58]. At 4.0 GPa there is a transition to the metallic phase II, with the monoclinic structure sm(4)[57]. This structure consists of puckered layers where each atom has 4 nearest neighbors. Phase III appears at 6.6 GPa, and is orthorhombic so(4), the result of shifting the monoclinic angle to 90°[57,58]. The I-II and II-III phase boundaries have been followed to triple points on the melting curve[59]. At 10.6 GPa, the β-Po rh(1) structure (IV) is found,

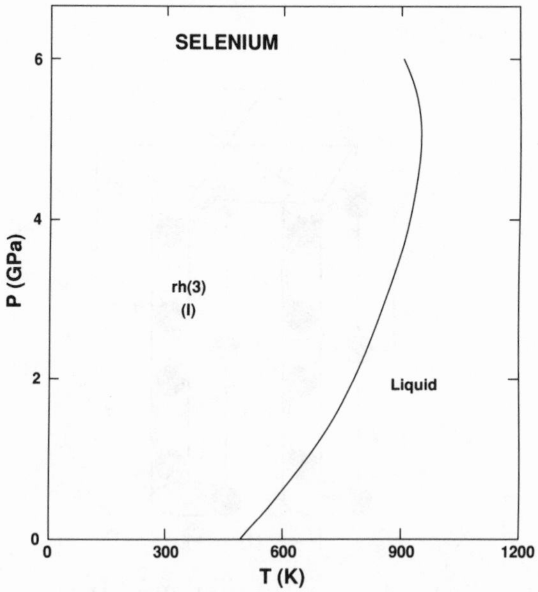

Fig. 10.6 The phase diagram of selenium.

which is a slightly distorted sc(1) lattice[56,58]. At 27 GPa, there is a transition to bcc, phase V. No further changes are found up to 38 GPa[58].

There has been rather little theoretical work on solid Te. AIP calculations on Te II indicate an anisotropic metal[60]. There has not yet been a serious attempt to predict the Te structures at high pressure.

The melting curve of Te has been measured to 6.0 GPa[50,59]. There is a pronounced maximum in the curve at 1.2 GPa and 740 K[61]. The negative slope of the melting curve above the maximum is reversed at the I-II-liquid triple point. The unusual shape of the melting curve together with the semiconductor-to-metal transition at low pressure have prompted experimental and theoretical work on liquid Te[62–65]. The emerging theoretical picture is of a liquid composed of helical chain fragments mixed with metallic Te atoms. Both temperature and pressure drive the dissociation of the chain fragments and increase the metallic component. There may be a pseudo-transition from semiconductor to metal in the liquid at 0.4 GPa on the melting curve, as indicated by resistance measurements and accurate melting measurements[61,62]. The Te phase diagram is shown in Fig. 10.7.

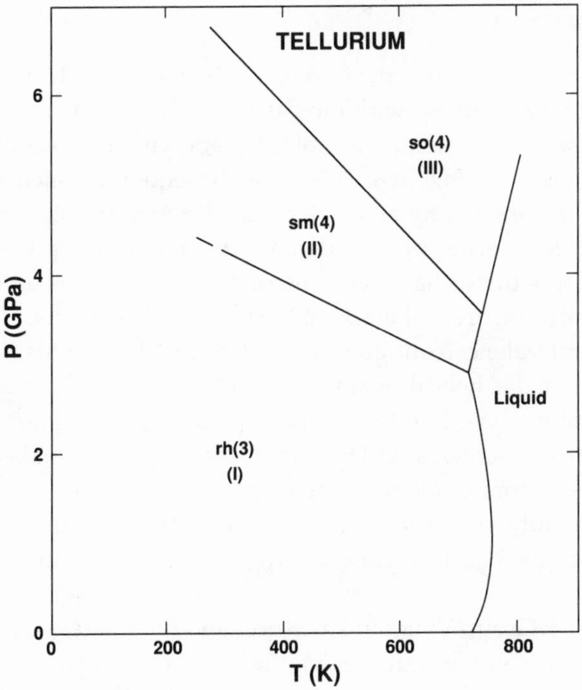

Fig. 10.7 The phase diagram of tellurium.

10.6 Polonium

Po is a radioactive element whose longest-lived isotope (209) has a half-life of 103 yr. The isotope commonly used for chemical study is 210, with a half-life of 138 days. At low temperatures, Po is a metal with the α sc(1) structure, the only element with this structure at RP[1]. A sluggish transition to the β rh(1) structure occurs between 291 and 327 K[66]. The β-Po rhombohedral angle is 98°, so this structure is only a slight distortion of the α structure. The density of the β phase is greater than that of the α phase, so that dP/dT for the transition will be negative. No high-pressure work has been reported for Po.

The sc(1) structure of Po has been studied with both nonrelativistic and fully relativistic tight-binding band-structure models[67]. The calculations show that the spin-orbit effect makes sc(1) Po more metallic by splitting the $6p$ band and suppressing the formation of covalent bonds. Thus it appears that relativistic effects favor the sc(1) structure of Po over the rh(3) structure characteristic of Te.

Po melts at 527 K. The melting curve has not been measured.

10.7 Discussion

Like the Group V elements, the Group VI elements shift from a diatomic molecular solid to a metal with increasing atomic number. The general progression seems to be diatomic molecule→polymeric molecular→open-lattice metal, shown in Fig. 10.8. This is also the sequence which occurs with increasing pressure on any given element, although the structures of the elements do not correspond in detail. A generalized phase diagram illustrating these trends has been constructed[68]. There is unfortunately a notable lack of theoretical work on the Group VI elements.

The general valence configuration of Group VI elements is s^2p^4. Diatomic O_2 and the helical chain forms of S, Se, and Te represent the dominance of sp^3-type bonding. Under pressure, the p character of the valence electrons increases, and this stabilizes simple-cubic-like structures where the neighbors lie along perpendicular axes corresponding to the p orbitals. Undoubtedly at very high pressures the Coulomb energy will stabilize more nearly close-packed structures, as found in Te V, which is bcc.

The complex Group VI liquid structures and the suggestion of separate liquid phases in S and Se indicate that the liquid phase of an element must not be thought of as featureless and less interesting than the solid phases. "Liquid-state physics" is just as important as "solid-state physics."

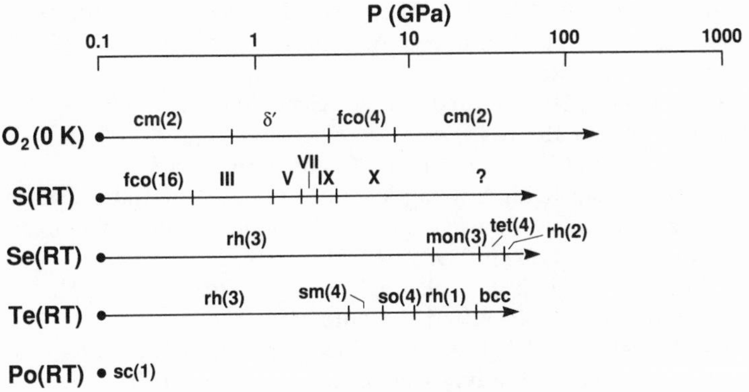

Fig. 10.8 Group VI structures at high pressure.

For Further Reading

B. Meyer, "Elemental Sulfur," Chem. Revs. **76**, 367 (1976).

References

1. J. Donohue, *The Structures of the Elements* (Wiley, New York, 1974) chap. 9.
2. R. J. Meier and R. B. Helmholdt, Phys. Rev. B **29**, 1387 (1984).
3. P. W. Stephens and C. F. Majkrzak, Phys. Rev. B **33**, 1 (1986).
4. F. Dunstetter, V. P. Plakhti, and J. Schweizer, J. Magn. Magn. Mat. **72**, 258 (1988).
5. R. Stevenson, J. Chem. Phys. **27**, 673 (1957).
6. R. J. Meier, M. P. van Albada, and A. Lagendijk, Phys. Rev. Lett. **52**, 1045 (1984).
7. H. J. Jodl, F. Bolduan, and H. D. Hochheimer, Phys. Rev. B **31**, 7376 (1985).
8. D. Schiferl, D. T. Cromer, L. A. Schwalbe, and R. L. Mills, Acta Cryst. **B39**, 153 (1983).
9. B. Olinger, R. L. Mills, and R. B. Roof, Jr., J. Chem. Phys. **81**, 5068 (1984).
10. D. Schiferl, personal communication.
11. M. Nicol and K. Syassen, Phys. Rev. B **28**, 1201 (1983).
12. S. Desgreniers, Y. K. Vohra, and A. L. Ruoff, J. Phys. Chem. **94**, 1117 (1990).
13. J. Yen and M. Nicol, J. Phys. Chem. **91**, 3336 (1987).
14. Yu. B. Gaididei and V. M. Loktev, Fiz. Nizk. Temp. **7**, 1305 (1981) [Sov. J. Low Temp. Phys. **7**, 634 (1981)].
15. C. A. English, J. A. Venables, and D. R. Salahub, Proc. Roy. Soc. Lond. A **340**, 81 (1974).
16. R. D. Etters, A. A. Helmy, and K. Kobashi, Phys. Rev. B **28**, 2166 (1983).
17. B. Kuchta and T. Luty, Chem. Phys. Lett. **126**, 506 (1986).
18. R. LeSar and R. D. Etters, Phys. Rev. B **37**, 5364 (1988).
19. A. P. J. Jansen and A. van der Avoird, J. Chem. Phys. **86**, 3583 (1987).
20. M. L. Klein, D. Levesque, and J.-J. Weis, Phys. Rev. B **21**, 5785 (1980).
21. R. D. Etters, K. Kobashi, and J. Belak, Phys. Rev. B **32**, 4097 (1985).

22. D. A. Young, C.-S. Zha, R. Boehler, J. Yen, M. Nicol, A. S. Zinn, D. Schiferl, S. Kinkead, R. C. Hanson, and D. A. Pinnick, Phys. Rev. B **35**, 5353 (1987).
23. B. Meyer, Chem. Revs. **76**, 367 (1976).
24. G. C. Vezzoli, F. Dachille, and R. Roy, Science **166**, 218 (1969).
25. S. Geller, Science **152**, 644 (1966).
26. G. C. Vezzoli and F. Dachille, Inorg. Chem. **9**, 1973 (1970).
27. L. Wang, Y. Zhao, R. Lu, Y. Meng, Y. Fan, H. Luo, Q. Cui, and G. Zou, in *High-Pressure Research in Mineral Physics*, M. H. Manghnani and Y. Syono, eds. (KTK Scientific, Tokyo, 1987) p. 299.
28. W. Häfner, H. Olijnyk, A. Wokaun, High-Press. Res. **3**, 248 (1990).
29. P. Wolf, B. Baer, M. Nicol, and H. Cynn, preprint.
30. L. C. Chhabildas and A. L. Ruoff, J. Chem. Phys. **66**, 983 (1977).
31. K. J. Dunn and F. P. Bundy, J. Chem. Phys. **67**, 5048 (1977).
32. M. J. Peanasky, C. W. Jurgensen, and H. G. Drickamer, J. Chem. Phys. **81**, 6407 (1984).
33. D. A. Golopentia and A. L. Ruoff, J. Appl. Phys. **52**, 1328 (1981).
34. H. G. David and S. D. Hamann, J. Chem. Phys. **28**, 1006 (1958).
35. S. S. Nabatov, A. N. Dremin, V. I. Postnov, and V. V. Yakushev, Pis´ma Zh. Eksp. Teor. Fiz. **29**, 407 (1979) [JETP Lett. **29**, 369 (1979)].
36. V. I. Postnov, L. A. Anan´eva, A. N. Dremin, S. S. Nabatov, and V. V. Yakushev, Fiz. Goreniya i Vzryva **22**, 106 (1986) [Combustion and Explosion **22**, 486 (1986)].
37. R. P. Rinaldi and G. S. Pawley, J. Phys. C **8**, 599 (1975).
38. C. M. Gramaccioli and G. Filippini, Chem. Phys. Lett. **108**, 585 (1984).
39. G. A. Saunders, Y. K. Yogurtçu, J. E. Macdonald, and G. S. Pawley, Proc. Roy. Soc. Lond. A **407**, 325 (1986).
40. G. C. Vezzoli, F. Dachille, and R. Roy, Inorg. Chem. **8**, 2658 (1969).
41. G. C. Vezzoli, F. Dachille, and R. Roy, J. Polymer Sci. A-1 **7**, 1557 (1969).
42. G. E. Sauer and L. B. Borst, Science **158**, 1567 (1967).
43. S. J. Kennedy and J. C. Wheeler, J. Chem. Phys. **78**, 1523 (1983).
44. B. M. Riggleman and H. G. Drickamer, J. Chem. Phys. **37**, 446 (1962).
45. A. R. Moodenbaugh, C. T. Wu, and R. Viswanathan, Solid State Comm. **13**, 1413 (1973).
46. F. P. Bundy and K. J. Dunn, J. Chem. Phys. **71**, 1550 (1979).
47. G. Parthasarathy and W. B. Holzapfel, Phys. Rev. B **38**, 10105 (1988).
48. D. Vanderbilt and J. D. Joannopoulos, Phys. Rev. B **27**, 6296 (1983).
49. H. Wendel, R. M. Martin, and D. J. Chadi, Phys. Rev. Lett. **38**, 656 (1977).
50. B. C. Deaton and F. A. Blum, Jr., Phys. Rev. **137**, A1131 (1965).
51. W. Klement, Jr., L. H. Cohen, and G. C. Kennedy, J. Phys. Chem. Solids **27**, 171 (1966).
52. I. E. Paukov, E. Yu. Tonkov, and D. S. Mirinskii, Russ. J. Phys. Chem. **41**, 995 (1967).
53. K. Tsuji, O. Shimomura, K. Tamura, and H. Endo, Z. Phys. Chem. NF **156**, 495 (1988).
54. W. W. Warren, Jr., and R. Dupree, Phys. Rev. B **22**, 2257 (1980).
55. V. V. Brazhkin, R. N. Voloshin, and S. V. Popova, High Press. Res. **4**, 348 (1990).
56. J. C. Jamieson and D. B. McWhan, J. Chem. Phys. **43**, 1149 (1965).

57. K. Aoki, O. Shimomura, and S. Minomura, J. Phys. Soc. Jpn. **48**, 551 (1980).
58. G. Parthasarathy and W. B. Holzapfel, Phys. Rev. B **37**, 8499 (1988).
59. F. A. Blum, Jr., and B. C. Deaton, Phys. Rev. **137**, A1410 (1965).
60. G. Doerre and J. D. Joannopoulos, Phys. Rev. Lett. **43**, 1040 (1979).
61. S. M. Stishov, N. A. Tikhomirova, and E. Yu. Tonkov, ZhETF Pis'ma **4**, 161 (1966) [JETP Lett. **4**, 111 (1966)].
62. E. Rapoport, J. Chem. Phys. **48**, 1433 (1968).
63. V. Ya. Prokhorenko, B. I. Sokolovskii, V. A. Alekseev, A. S. Basin, S. V. Stankus, and V. M. Sklyarchuk, Phys. Stat. Sol. (b) **113**, 453 (1982).
64. S. Takeda, S. Tamaki, and Y. Waseda, J. Phys. Soc. Jpn. **53**, 3830 (1984).
65. J. Hafner, J. Phys: Condens. Matter **2**, 1271 (1990).
66. J. M. Goode, J. Chem. Phys. **26**, 1269 (1957).
67. L. L. Lohr, Inorg. Chem. **26**, 2005 (1987).
68. G. C. Vezzoli, L. W. Doremus, and P. J. Walsh, Phys. Stat. Sol. (a) **32**, 683 (1975).

CHAPTER 11
The Group VII Elements (The Halogens)

11.1 Introduction

The halogens are all diatomic molecules, with the possible exception of At, which has never been isolated as a pure element. The halogens carry to the limit the trend toward covalency in moving from left to right across the periodic table. Of great current interest is the experimentally achieved high-pressure metallization and dissociation of I_2, which serves as a model for the high-pressure metallization of other diatomic solids.

11.2 Fluorine

Below 45.6 K at RP, α-F_2 is monoclinic with 4 molecules per unit cell, cm(4)[1]. The precise space group is uncertain; it is either C2/m or C2/c. These structures are both very similar to α-O_2 and they differ in the tilt of the molecular axis away from the normal to the ab plane. Above 45.6 K, β-F_2 is sc(8), isomorphic with γ-O_2[1].

Because F_2 is chemically very reactive, it is difficult to study at high pressure. Nevertheless, DAC Raman measurements have determined the α–β phase boundary up to 5 GPa[2]. No new phases are found up to 6 GPa[2].

Theoretical work on solid F_2 has focused on the search for an intermolecular potential which can reproduce the observed lattice properties[3,4]. Calculations have been performed with simple atom-atom pair potentials plus electric quadrupoles. The model potentials can be fitted to yield adequate agreement with static and dynamic data for the solid. The calculations indicate that the C2/c and C2/m structures are indeed very close in energy and that the more stable of the two cannot be confidently predicted with simple potential models. The quadrupole moment of F_2 lies between that of N_2 and O_2, and the solid structure of α-F_2 is a compromise between the α-N_2 and β-O_2 structures[4].

F_2 melts at 53.5 K. The melting curve has been determined with rather low accuracy to 2.5 GPa[2]. Although the RP melting temperature is very close to that of O_2, the slope of the curve is much closer to that of N_2[5]. The F_2 phase diagram is shown in Fig. 11.1.

11.3 Chlorine

Solid Cl_2 is orthorhombic Cmca, eco(4), with 4 molecules per unit cell[1]. The molecules are packed in layers parallel to the bc plane, with the orientations of the layer molecules alternating in a zigzag pattern. Compression to 45 GPa shows continuous changes in the orthorhombic lattice parameters and Raman frequencies, but no phase transition[6,7].

Theoretical work on solid Cl_2 has focused on the correct form of the intermolecular potential required to predict the observed crystal structure[8,9]. It has become clear that anisotropic dispersion and repulsion

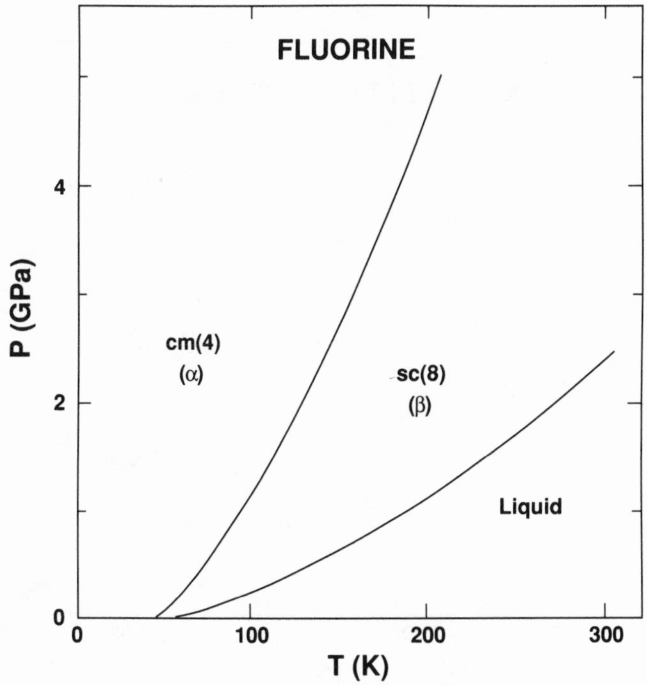

Fig. 11.1 The phase diagram of fluorine.

forces as well as higher-order electric multipoles are needed to mimic the highly nonspherical electron density found in Cl_2. An adequate fit to the Cmca cell parameters and lattice-vibration frequencies has been obtained.

The melting curve of Cl_2 has been measured to 0.7 GPa[10]. The Cl_2 phase diagram is shown in Fig. 11.2.

11.4 Bromine

Solid Br_2 is eco(4), isomorphic with Cl_2[1]. Br_2 has been compressed at RT up to 88 GPa[6,11]. At ca. 80 GPa there is a transition to the bco(2) monatomic phase also seen in I_2[11]. Reflectivity measurements to 170 GPa show strong metallic behavior above 100 GPa[12].

The Br_2 intermolecular potential has been modeled with atom-atom and higher electric-multipole terms, and the result fits static and dynamic data adequately, favoring the Cmca over the Pa3 structure[13,14].

The melting curve of Br_2 has been measured to 1.0 GPa[10,15]. The Br_2 phase diagram is shown in Fig. 11.2.

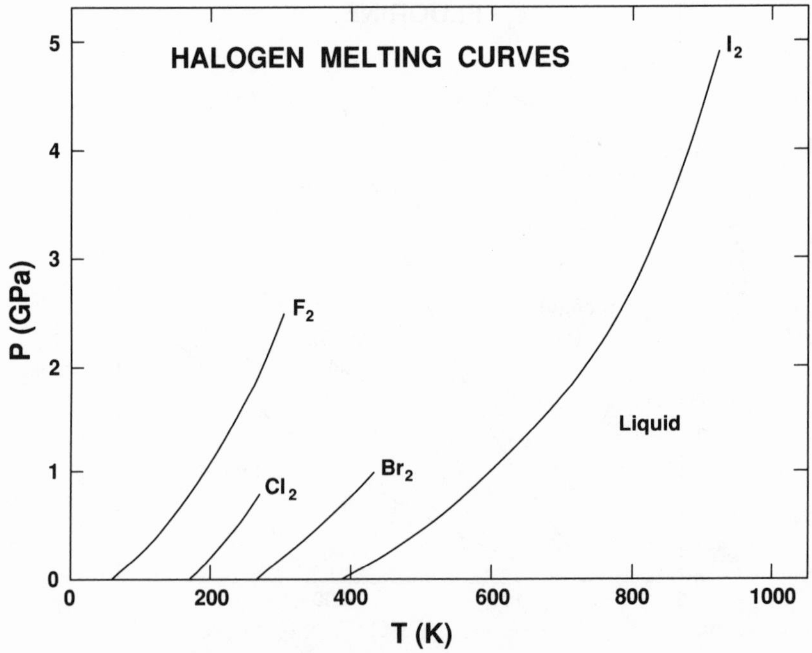

Fig. 11.2 The melting curves of the halogens.

11.5 Iodine

At RTP, I_2 is eco(4), isomorphic with Cl_2 and Br_2[1]. RT compression produces a very interesting series of phase transitions. As the pressure is increased, the molecules within the bc layer planes approach each other while the intramolecular distance remains constant[16]. At 17.0 GPa, I_2 becomes a diatomic metal with strongly anisotropic resistance[17]. This band-overlap metallization is apparently a higher-order phase transition showing no volume change. At 21.0 GPa, the interatomic distances between molecules and within molecules become equal and a first-order phase transition leads to a bco(2) monatomic metal[18,19]. Although Mössbauer measurements on I_2 at low temperatures suggest the continuation of diatomic bonding above 21.0 GPa[20], XRD experiments at 35 K confirm the appearance of the dissociated bco(2) phase at ca. 21.5 GPa[21]. This sequence of crystal-structure changes is shown in Fig. 11.3. Further compression to 45 GPa causes symmetrization to a ct(2) lattice, where $a = c \neq b$[22]. This may be a second-order transition. Finally, another transition at 55 GPa yields a completely isotropic fcc lattice. No further

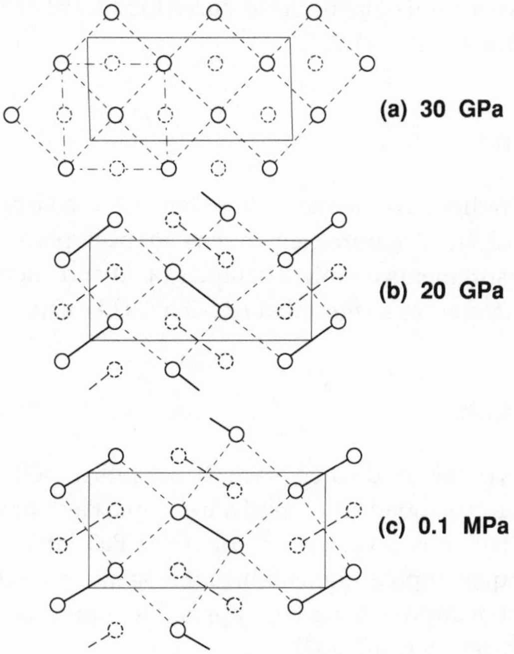

Fig. 11.3 The crystal structure changes in I_2 under pressure. (From Takemura, et al.[18]. Redrawn with permission.)

transitions are seen up to 180 GPa[12,23]. Shock compression of I_2 to 170 GPa shows a Hugoniot which is consistent with a monatomic metallic state[24].

Several pair-potential models of solid molecular I_2 have been developed[14,25]. It is evident from the interatomic distances in the molecular crystal that charge transfer is important even at RP[14,25]. An intermolecular potential model including a charge-transfer term has been fitted to high-pressure structural data and Raman spectra. Potential parameters show a residual volume dependence which is ascribed to an inadequate charge-transfer description[25]. An alternative potential which ignores charge transfer has been fitted to RP data with modest success[14].

There has been rather little theoretical work on the insulator-metal transition in I_2. Very simple LCAO Hückel calculations on I_2 under pressure show a band closure at ca. 15 GPa, in surprisingly good agreement with experiment[26]. For the monatomic phase, there have been a few band-structure calculations[24,27,28]. Comparison of the densities of states of ct(2) and fcc I computed for 49 GPa show that ct(2) will have the lower energy, as observed, and that further compression will reverse this relationship[28].

The melting curve of I_2 has been measured to 5.0 GPa[15,29]. The I_2 phase diagram is shown in Fig. 11.2.

11.6 Astatine

At is a highly radioactive element whose longest-lived isotope is 210, with a half-life of 8.3 hr. The pure element has not been prepared in bulk, but volatility measurements on micro-samples of At and chemical studies on the aqueous At^- species suggest that At behaves like a normal halogen[30].

11.7 Discussion

The structures of the solid halogens at RP are rather well understood as a balancing of electric-quadrupole and anisotropic Pauli forces[31–33]. The eco(4) Cmca lattice is favored over the sc(4) Pa3 lattice because of the relatively low quadrupole moments in the halogens. A similar intermolecular pair-potential approach has been applied to liquid-halogen structures, also with satisfactory results[34].

The high-pressure behavior of the solid halogens is summarized in Fig. 11.4. The isotherms may be compared to one another by scaling the lattice parameters with the bond length in the solid state[11]. In Fig. 11.5, a plot of scaled volume V^* vs. pressure for Br_2, IBr, and I_2 shows that dissociation to the monatomic metal occurs at a constant value $V^* = 1.29$. Chlorine is expected to dissociate at pressures well above 100 GPa[35]. The sequence of transitions observed in I_2 from anisotropic molecular solid to diatomic metal to isotropic monatomic metal provides a useful model for how other diatomics might behave at higher pressure.

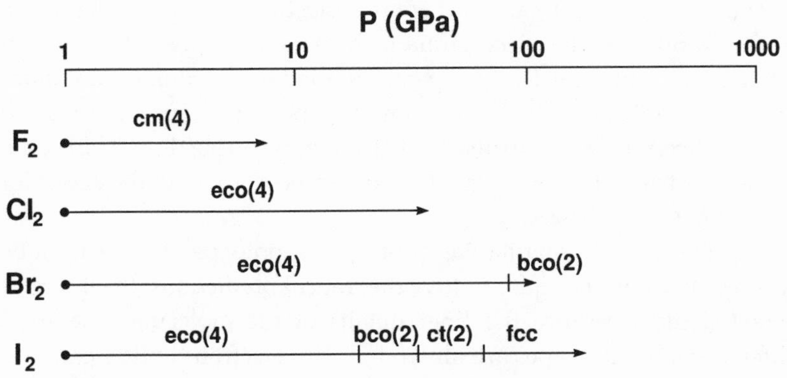

Fig. 11.4 The 0 K structures of the halogens at high pressures.

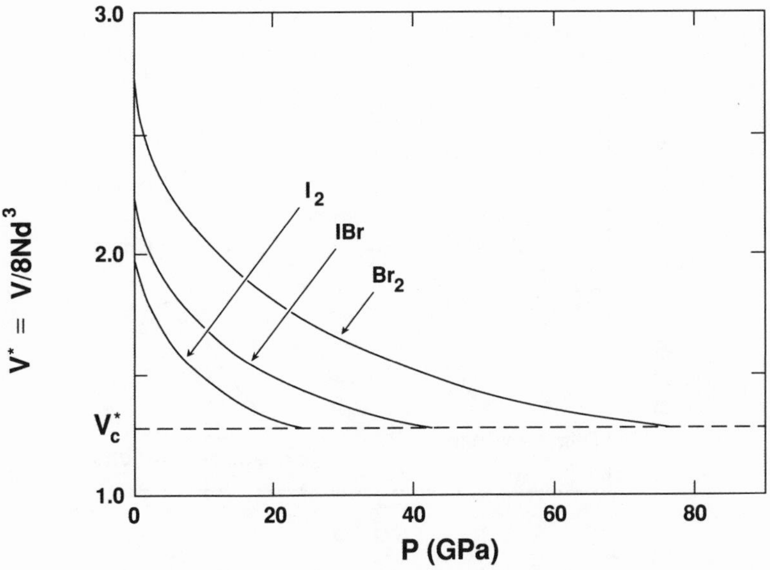

Fig. 11.5 The scaled volumes vs. pressure for Br_2, IBr, and I_2.

It is appropriate here to discuss the general problem of diatomic phase diagrams. The phase diagrams of the diatomic elements H_2, N_2, O_2, F_2, Cl_2, Br_2, and I_2 have been studied under (P,T) conditions which reveal the stability ranges of the most common diatomic phases. It is clear that these phases are close approximations to fcc and hcp close packing and that the repulsive intermolecular forces dominate the crystal structures. The nonspherical shapes of the molecules allow for the numerous exotic space groups which are variations on the theme of close packing.

The two dominant variables controlling the structural stability of diatomic solids are the electric-quadrupole moment and the elongation, defined as the internuclear distance d divided by the atom-atom diameter σ[31]. A simple Lennard-Jones atom-atom potential plus a point quadrupole predicts the $T=0, P=0$ phase diagram shown in Fig. 11.6. The predicted phases are in fact the experimentally observed ones, with the exception of the magnetic O_2 phases.

Fig. 11.6 is only qualitative because the potential model is only an approximate one. For quantitative theoretical predictions, the interatomic potential must capture the finer details of the molecular shape. The different diatomic shapes are illustrated with electron-density contours in Fig. 11.7[36]. The subtle differences in shape shown in Fig. 11.7 require potential models much more elaborate than the simple atom-atom model[32]. Such models have been used to predict accurately the 0 K high-pressure

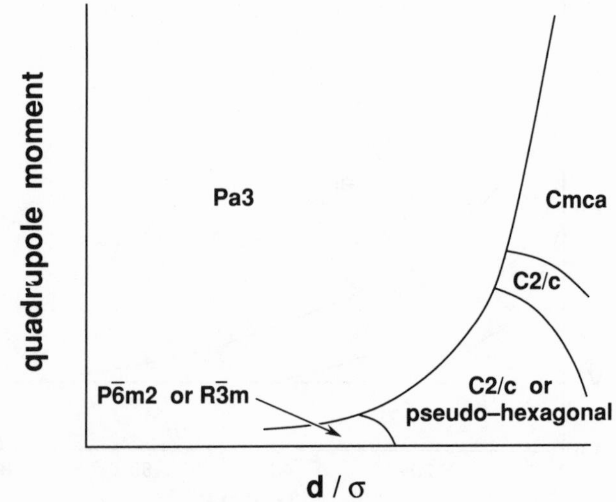

Fig. 11.6 General diatomic phase diagram. (From English and Venables[31]. Redrawn with permission.)

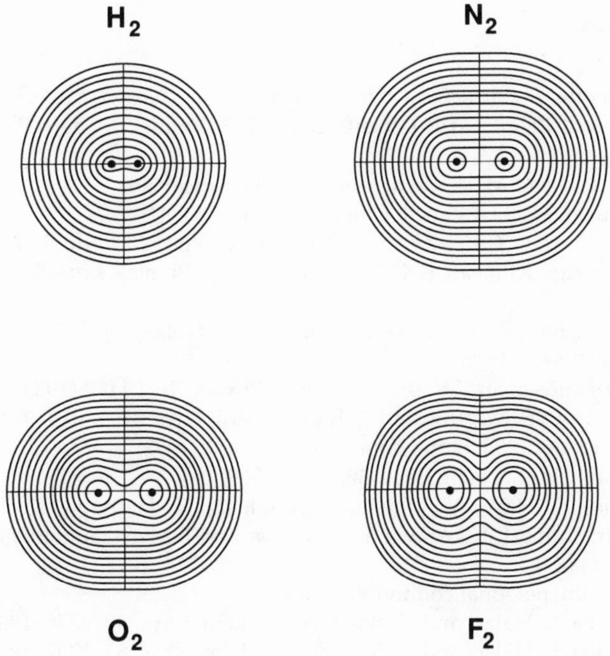

Fig. 11.7 Electron-density contours of diatomics. (From Wahl[36]. © 1966 by the AAAS. Redrawn with permission.)

phases in N_2 and O_2, as described in chapters 9 and 10. So far, simple statistical-mechanical models have not been successful in predicting the higher-temperature diatomic phases in which rotational motions are important. It now appears that MC and MD simulations are the appropriate tools for determining the diatomic temperature-driven phase transitions, including melting.

There is intense interest in the mechanisms of metallization and pressure dissociation of the diatomic solids. With the DAC, rapid progress has been made toward understanding these processes. The clearest case of metallization so far is I_2, which metallizes by band overlap at 17.0 GPa. Dissociation then occurs in stages by symmetrization of the crystal structure. Two other diatomics, O_2 and Br_2, have also been metallized by band overlap. Whether H_2 and N_2 have been metallized remains to be proven. Theoretical predictions of metallization pressures are promising, but are hampered by lack of knowledge of the correct crystal structures and by the limitations of the local-density approximation.

References

1. J. Donohue, *The Structures of the Elements* (Wiley, New York, 1974) chap. 10.
2. D. Schiferl, S. Kinkead, R. C. Hanson, and D. A. Pinnick, J. Chem. Phys. **87**, 3016 (1987).
3. K. Kobashi and M. L. Klein, Mol. Phys. **41**, 679 (1980).
4. D. Kirin and R. D. Etters, J. Chem. Phys. **84**, 3439 (1986).
5. D. A. Young, C.-S. Zha, R. Boehler, J. Yen, M. Nicol, A. S. Zinn, D. Schiferl, S. Kinkead, R. C. Hanson, and D. A. Pinnick, Phys. Rev. B **35**, 5353 (1987).
6. E.-Fr. Düsing, W. A. Grosshans, and W. B. Holzapfel, J. Phys. (Paris) **45**, C8-203 (1984).
7. P. G. Johannsen and W. B. Holzapfel, J. Phys. C **16**, L1177 (1983).
8. E. Burgos, C. S. Murthy, and R. Righini, Mol. Phys. **47**, 1391 (1982).
9. S. L. Price and A. J. Stone, Mol. Phys. **47**, 1457 (1982).
10. S. E. Babb, Jr., J. Chem. Phys. **50**, 5271 (1969).
11. Y. Fujii, K. Hase, Y. Ohishi, H. Fujihisa, N. Hamaya, K. Takemura, O. Shimomura, T. Kikegawa, Y. Amemiya, and T. Matsushita,Phys. Rev. Lett. **63**, 536 (1989).
12. R. Reichlin, personal communication.
13. Z. Gamba, E. Halac, and H. Bonadeo, J. Chem. Phys. **80**, 2756 (1984).
14. Z. Gamba, E. Halac, and H. Bonadeo, J. Chem. Phys. **85**, 1202 (1986).
15. I. E. Paukov, E. Yu. Tonkov, and D. S. Mirinskii, Russ. J. Phys. Chem. **41**, 995 (1967).
16. O. Shimomura, K. Takemura, Y. Fujii, S. Minomura, M. Mori, Y. Noda, and Y. Yamada, Phys. Rev. B **18**, 715 (1978).
17. N. Sakai, K. Takemura, and K. Tsuji, J. Phys. Soc. Jpn. **51**, 1811 (1982).
18. K. Takemura, S. Minomura, O. Shimomura, and Y. Fujii, Phys. Rev. Lett. **45**, 1881 (1980).
19. K. Takemura, S. Minomura, O. Shimomura, Y. Fujii, and J. D. Axe, Phys. Rev. B **26**, 998 (1982).
20. M. Pasternak, J. N. Farrell, and R. D. Taylor, Phys. Rev. Lett. **58**, 575 (1987).
21. H. Fujihisa, Y. Fujii, K. Hase, Y. Ohishi, N. Hamaya, K. Tsuji, K. Takemura, O. Shimomura, H. Takahashi, and T. Nakajima, High-Press. Res. **4**, 330 (1990).
22. Y. Fujii, K. Hase, Y. Ohishi, N. Hamaya, and A. Onodera, Solid State Comm. **59**, 85 (1986).
23. Y. Fujii, K. Hase, N. Hamaya, Y. Ohishi, A. Onodera, O. Shimomura, and K. Takemura, Phys. Rev. Lett. **58**, 796 (1987).
24. A. K. McMahan, B. L. Hord, and M. Ross, Phys. Rev. B **15**, 726 (1977).
25. K. Kobashi and R. D. Etters, J. Chem. Phys. **79**, 3018 (1983).
26. F. Siringo, R. Pucci, and N. H. March, Phys. Rev. B **38**, 9567(1988).
27. Y. Natsume, in *Solid State Physics under Pressure*, S. Minomura, ed. (KTK Scientific, Tokyo, 1985) p. 43.
28. N. Orita, T. Sasaki, and K. Niizeki, Solid State Comm. **64**, 391 (1987).
29. W. Klement, Jr., L. H. Cohen, and G. C. Kennedy, J. Chem. Phys. **44**, 3697 (1966).
30. E. Anders, Ann. Rev. Nucl. Sci. **9**, 203 (1959).

31. C. A. English and J. A. Venables, Proc. Roy. Soc. Lond. A **340**, 57 (1974).
32. S. L. Price, Mol. Phys. **62**, 45 (1987).
33. V. V. Nauchitel´ and I. B. Golovanov, Kristallografiya **32**, 1347 (1987) [Sov. Phys. Cryst. **32**, 791 (1987)].
34. M. Misawa, J. Chem. Phys. **91**, 2575 (1989).
35. P. G. Johannsen, E. F. Düsing, and W. B. Holzapfel, in Ref. 27, p. 105.
36. A. C. Wahl, Science **151**, 96 (1966).

CHAPTER 12
The Group VIII Elements
(The Rare Gases)

12.1 Introduction

Because the rare gases are monatomic and appear to be well-characterized by pairwise-additive interaction potentials, their thermodynamic properties and phase diagrams have stimulated a large literature on the development of semiempirical potential functions and statistical-mechanical theories of liquids and solids. Much of the advance of liquid theory can be attributed to the attempt to predict accurately the properties of liquid argon. That the effective potential functions for the rare gases all have approximately the same form has given rise to the corresponding-states theory, in which the phase diagrams can be described by a common function of dimensionless variables. In the case of helium, the quantum effects are so large as to dominate the thermodynamic properties near 0 K, and from these observations the theory of quantum liquids and solids has arisen.

In many ways the phase diagrams of the rare gases can be said to be well understood, but there remain subtle features of these diagrams which have so far escaped explanation. Continuing research is thus called for.

12.2 Helium

Helium is the only element that is not solid at 0 K and RP. This is readily explained in terms of its small mass and weak intermolecular attraction, leading to very-large-amplitude zero-point motion. This motion is sufficient to disrupt any lattice, and the ground state is therefore a fluid. Near 0 K, the two isotopes He-3 and He-4 show strong quantum-liquid behavior, which has stimulated an immense experimental and theoretical literature[1]. However, since the overall form of the helium-isotope phase diagrams is not strongly influenced by the details of the low-pressure quantum-fluid properties, only the briefest description of these properties will be given here. The main difference between the two isotopes is that He-3 is a fermion, while He-4 is a boson.

He-3 behaves as a normal Fermi liquid down to a temperature of a few mK, where transitions to the A or B magnetic superfluids occur[2,3]. The very low-temperature phase diagram is shown in Fig. 12.1. The transition from A or B to the normal Fermi liquid is second order, but the A-B transition is first order, the only known example of such a transition in a liquid element. At 3.4 MPa, these fluid phases freeze to the bcc solid. The melting curve shows a noticeable pressure minimum at 0.32 K[4], as seen in Fig. 12.2.

In He-4, the normal fluid (I) undergoes a λ-transition to a superfluid phase (II) at 2.17 K and RP. The I-II phase boundary has a negative dP/dT. He-4 at 0 K freezes to an hcp solid at 2.5 MPa. There is a bcc phase occupying a very small (P,T) region near the melting curve, and a very weak pressure

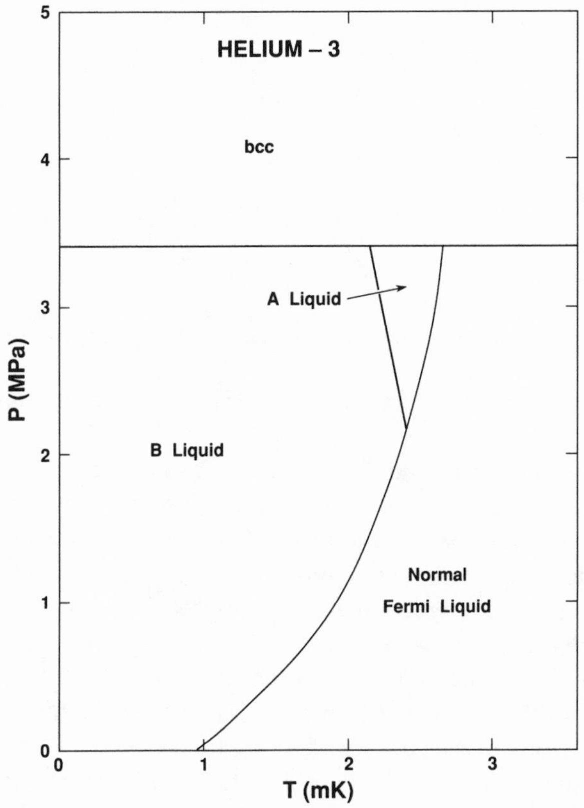

Fig. 12.1 The helium-3 phase diagram at very low temperature.

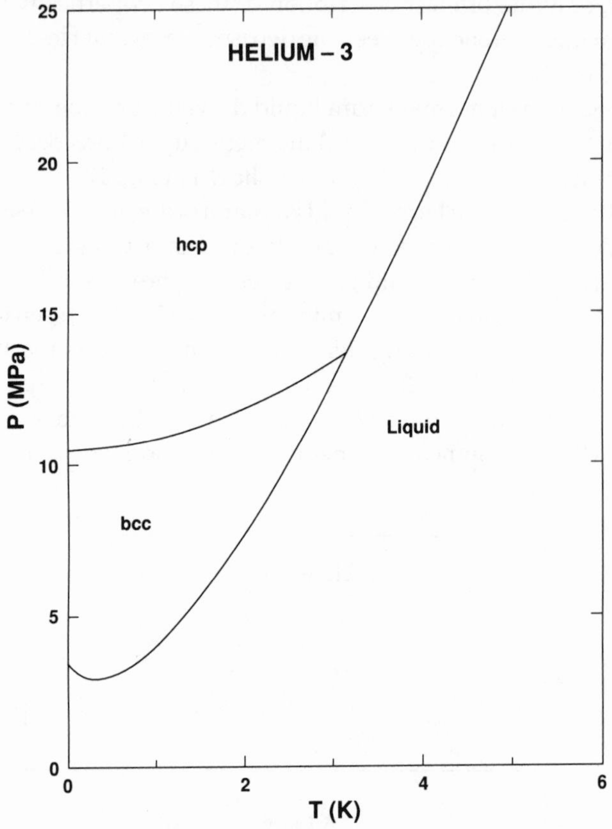

Fig. 12.2 The helium-3 phase diagram at low temperature.

minimum in the melting curve near 0.8 K[5]. These features are shown in Fig. 12.3.

The theory of quantum liquids developed by Landau postulates that an interacting system of particles near 0 K can be described in terms of a gas of quantized excitations or quasiparticles[6]. As the temperature rises, these quasiparticles begin to interact, and at some point the description loses validity because the interaction energies approach the quasiparticle energies themselves.

In He-3, the Fermi liquid is well described by the Landau theory from ca. 0.1 K down to the superfluid transitions[7]. The superfluid transition is closely analogous to the superconducting transition, and arises from the pairing of nuclear spins. The paired state is a triplet, and the A and B superfluids represent different triplet spin orderings. The Fermi liquid-superfluid transition is of the Bardeen-Cooper-Schrieffer type[2,8].

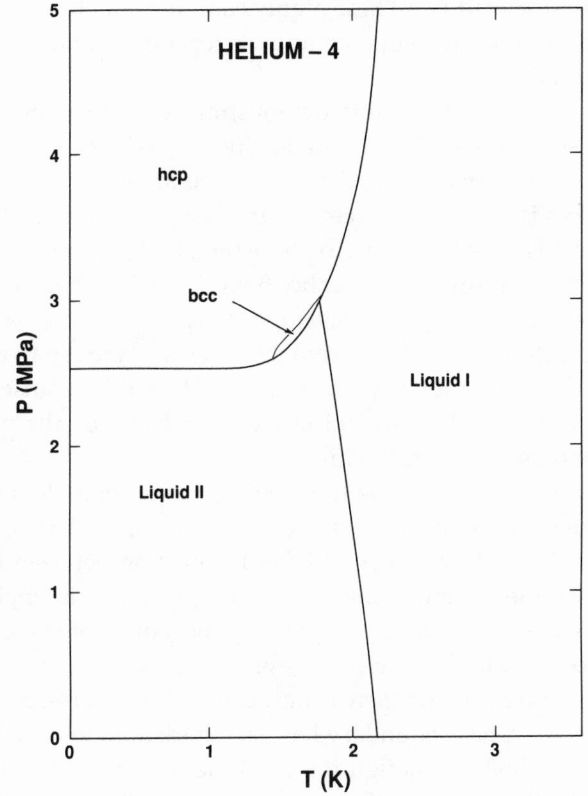

Fig. 12.3 The helium-4 phase diagram at low temperature.

Many details of the superfluid He-3 phases remain to be worked out, but it is a triumph of low-temperature physics that the complete phase diagram together with an elaborate theoretical structure were constructed within a few years of the original discovery of the superfluids in 1972[2,3].

The Landau theory for He II postulates two types of excitations: phonons and rotons. This description works up to ca. 1.5 K, and cannot describe the λ-transition. An ideal Bose gas with a particle mass equal to that of He-4 shows a Bose-Einstein transition at 3.14 K, and it is clear that there must be a connection between the λ-transition at 2.17 K and the Bose-Einstein transition[9]. This connection was difficult to make until now because of computational limitations. However, recent Monte Carlo path-integral calculations for He-4 modeled as bosons interacting with the Aziz pair potential now convincingly predict the λ-transition and the presence of the zero-momentum condensate below the transition[10]. The Bose condensate is predicted to be only 10% of the fluid at $T = 0$ K, rather than 100% as found in the ideal Bose gas[10,11]. These highly computer-intensive calculations illustrate the importance of supercomputers for solving problems of many-body quantum theory.

Variational QMC calculations demonstrate that the ground state of He-3 and He-4 at 0 K is the liquid, and that freezing is indeed expected in the observed pressure range[11,12]. Also, the calculations have shown that at 0 K liquid He-3 is expected to freeze to the bcc phase (Fig. 12.2), and that He-4 is expected to freeze to the hcp phase (Fig. 12.3), as observed[13]. The higher freezing pressure and larger bcc field of He-3 are both ascribable to the larger zero-point energy of this isotope. The pressure minima in the low-temperature melting curves arise from the reversal of the usual rule that the liquid has a higher entropy than the solid. In He-3, this is caused by more spin disorder in the solid[14], and in He-4, probably by the presence of transverse phonons in the solid[15].

At approximately 0.1 GPa on the melting curve, solid He undergoes a transition from hcp to fcc. In recent years this transition has been studied in great detail[16–18]. As expected for a transition between two close-packed phases, the volume change across the transition is small, $\Delta V/V \approx 10^{-5}$. The transition is martensitic, involving the motion of planes of atoms, which induces a considerable hysteresis and makes the location of the equilibrium phase boundary increasingly difficult as the pressure increases. So far, the hcp-fcc phase boundary has been measured to 0.9 GPa in He-4 and 0.6 GPa in He-3. Although the He-3 melting curve lies at a higher pressure[19,20] than in He-4, the solid-solid phase lines have the reverse relationship, as shown in Fig. 12.4.

With the DAC, the melting curve of He-4 has been measured up to 460 K and 24 GPa[21-23]. The earlier workers found an anomaly on the melting curve which might represent a triple point[21,22], but the latest measurements find no anomaly[23]. Recent XRD measurements with synchrotron radiation show that solid He at RT is hcp up to 23.3 GPa, which suggests that the fcc phase has disappeared in a triple point on the melting curve at some lower pressure[24]. The high-pressure He-4 melting curve is shown in Fig. 12.5.

The most complete attempt to develop a theory of the hcp-fcc transition is based on the observation that the quantum ground state of a particle in an hcp cell is lower than that in an fcc cell[25]. This explains the stability of hcp He at very low temperatures. With increasing temperature, lattice vibrations, represented by harmonic lattice dynamics, stabilize the fcc phase. This theory adequately predicts the observed transition temperature, and in addition predicts that the phase line must eventually bend back toward the

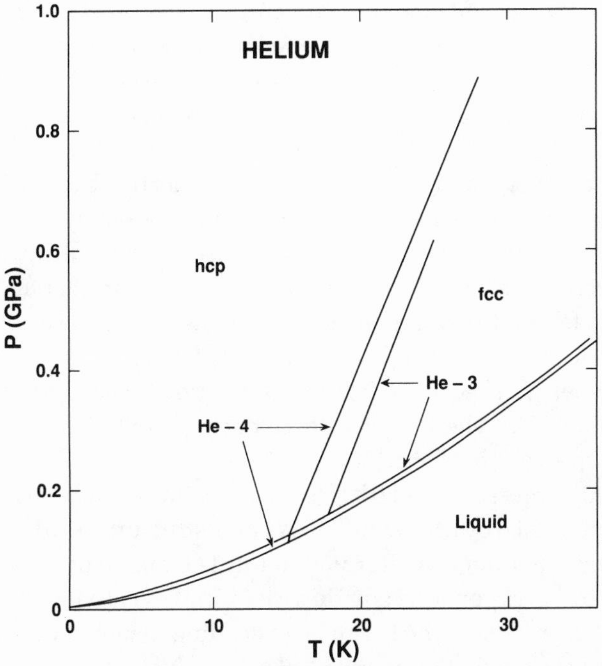

Fig. 12.4 Melting curves and hcp-fcc transitions in He-3 and He-4.

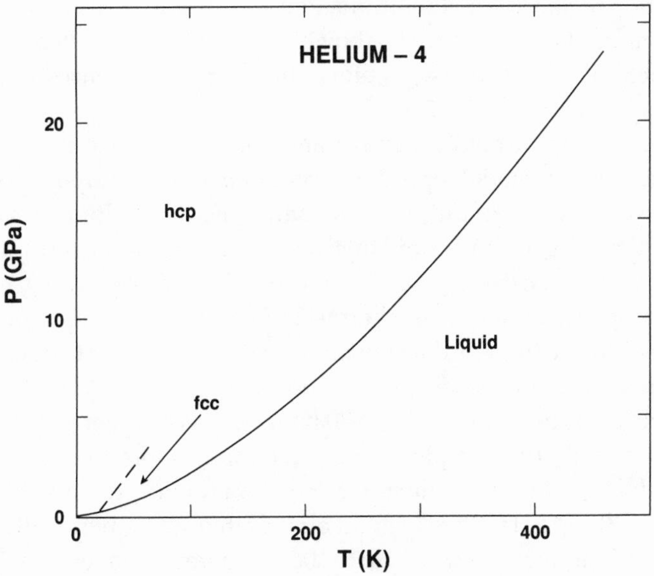

Fig. 12.5 The helium-4 phase diagram at high pressure.

0 K axis, since zero-point motion will eventually stabilize fcc. Similar results are obtained with a three-body potential and the self-consistent phonon model[26]. However, unless there are two separate hcp phases, this theory is inconsistent with the observation that hcp is present at high pressure and RT[24].

Theoretical calculations of various types predict that bcc will be stable at high pressure and temperature not far from the anomaly on the melting curve[27–29]. The bcc phase results from the softening of the pair potential with the decrease in particle separation, as discussed in chapter 3. The bcc structure is favored over fcc at high temperature because of its higher entropy. At low temperature, bcc is ultimately favored by virtue of its lower zero-point energy. Theory predicts a 0 K transition to bcc at pressures above 100 GPa[26,28,29]. The use of a three-body potential predicts the 0 K sequence hcp→fcc→hcp→bcc[26].

At very high pressures and 0 K, He must become metallic. This pressure has been determined by LMTO and ASW band-structure calculations[30,31]. Low-pressure He is an insulator with a full 1*s* band below a large energy gap. With compression, this gap decreases, and eventually the bottom of the 2*p* band drops below the Fermi level, leading to metallic behavior. The volume, and especially the pressure, at which this occurs depends on the crystal structure. For bcc, fcc, and hcp, LMTO metallization pressures of 3.15, 9.7, and 11.2 TPa are found[30]. Since hcp has the lowest energy in the metallization region, the metallic transition pressure at 0 K is 11.2 TPa. ASW calculations predict 11.0 TPa[31]. Above 100 TPa, He starts to become OCP-like and bcc is the stable phase. A plot of the energy differences among the He phases is shown in Fig. 12.6.

In order to compute high-pressure phase transitions in He, it is necessary that the free-energy model reproduce accurately the measured equation of state in the (P,T) region of the phase transition. For He melting, it is natural to assume that the solid and liquid thermodynamic properties are governed by a simple pair potential, which may be obtained by fitting existing equation of state data, such as isotherms[24,32] and shock Hugoniots[33]. It is in fact possible to fit these data with a simple potential and to obtain good agreement with the experimental melting curve[29,30,34]. The melting curve of He-3 has been measured to 30 K, and it has a higher pressure than the He-4 curve[20]. Simple theoretical models predict the reverse relationship[35]. New path-integral calculations of the pressure difference $\Delta P = P_m(\text{He-3}) - P_m(\text{He-4})$ are in good quantitative agreement with experiment and they predict that $\Delta P > 0$ up to 100 K, where it changes sign[36]. For

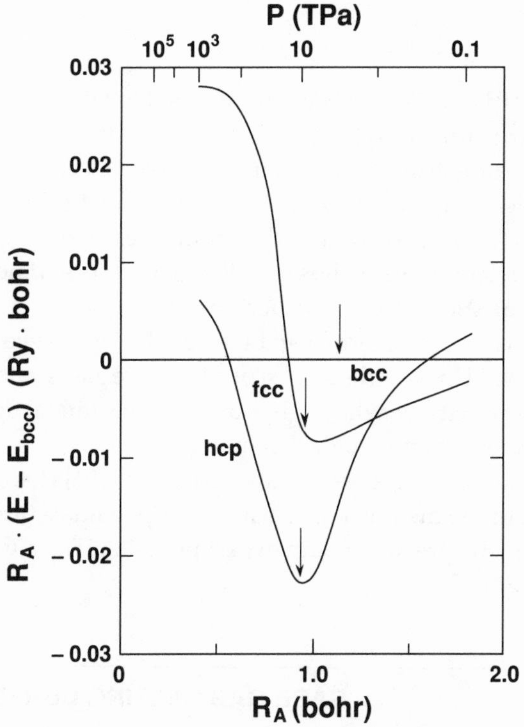

Fig. 12.6 Comparison of LMTO energies for helium. Metallization points are indicated by arrows.

$T > 100$ K, $\Delta P < 0$, in qualitative agreement with the predictions of the theoretical models based on the Wigner-Kirkwood quantum correction[34].

An interesting result of such calculations is a significant discrepancy between the best *ab initio* or experimental two-body potential[37–39] and the fitted potential. The fitted potential is lower, indicating the existence of attractive many-body interactions in dense He[24,30]. For melting predictions, these many-body terms may be absorbed into an effective two-body potential, but for solid-solid transitions, the phase transitions will be sensitive to the exact form of the many-body terms[26].

The overall phase diagram of He is surprisingly complex, being determined by harmonic and anharmonic atomic motions, and by changes in band structure under compression. The tenfold volume compression achieved over the accessible pressure range extends from the quantum fluid to the nearly classical harmonic solid. There is as yet no single theory which covers this range, and so the helium phase diagram remains an outstanding challenge for physics.

12.3 Neon

Solid Ne at RP is fcc. Ne has been compressed to 110 GPa at RT without any change of structure[40–42].

Electron-band-structure calculations show a very high metallization pressure, 134 TPa according to LCGTO calculations[43], and 158 TPa according to APW calculations[44], in both cases assuming an fcc structure. Since the heavier rare-gas solids metallize at pressures much lower than for He (11.2 TPa), Ne is "out of order" in the sequence of pressures. The explanation is that the energy gap between the $2p$ valence band and the $3d$ conduction band in Ne is the largest of all the rare gases, and that in addition Ne has core electrons which give a repulsive interaction and hence a pressure at metallization higher than in He.

The melting curve has been measured to 5.5 GPa[41,45,46]. Theoretical calculations of the melting curve using pair potentials fitted to isothermal data are in good agreement with experiment[46]. The Ne melting curve is shown in Fig. 12.7.

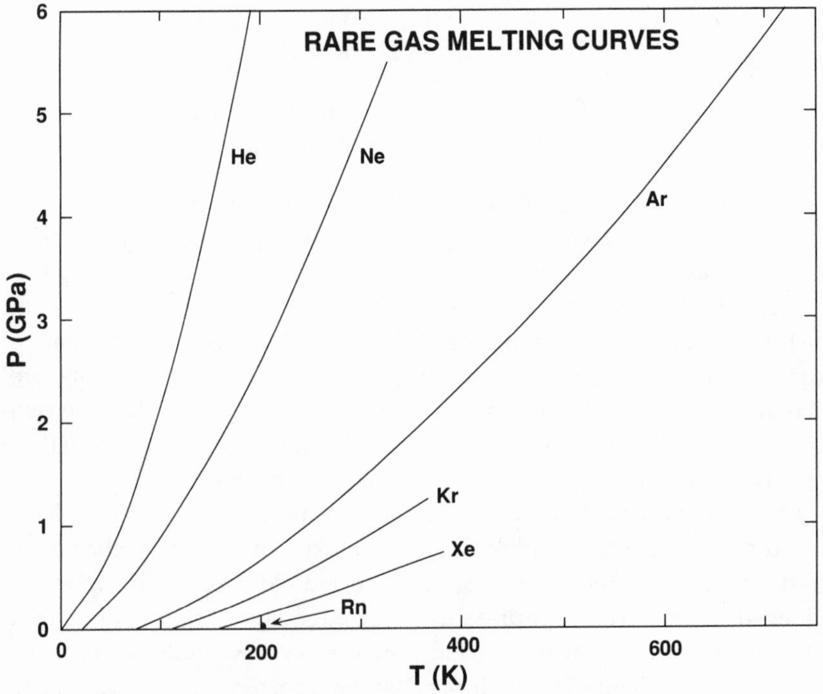

Fig. 12.7 The rare-gas melting curves.

12.4 Argon

Because of its abundance and ease of handling, Ar has been very intensively studied, and it has become the paradigm of the simple classical liquid or solid. The equation of state is known accurately over a very broad range of (*P*,*T*) conditions, and innumerable theoretical papers have been written on the Ar intermolecular potential and on statistical-mechanical models of the solid and liquid states of Ar[47].

Solid Ar at RP is fcc. Ar has been compressed in the DAC at RT to 80 GPa with no change in structure[41,48]. Extensive shock data are also available, including pyrometric measurements[49–51].

LMTO calculations predict that compression leads to the sequence of structures fcc→hcp→bcc→fcc[52]. The initial fcc-hcp transition pressure cannot be determined precisely because of the extremely small energy difference, but it is below 230 GPa. The hcp-bcc and bcc-fcc transitions are expected at 0.97 and 2.2 TPa, respectively. Metallization of Ar is predicted to occur by the crossing of the 3*p* and 3*d* bands at 430 GPa in the hcp phase[52]. This has not been directly observed experimentally, but the pressure-volume shock Hugoniot shows an anomalous "softening" or flattening due to excitation of electrons across the decreasing band gap[53]. Theoretical Hugoniot calculations using energy gaps computed by band-structure models are in good agreement with experiment, thus supporting the metallization prediction[53].

The melting curve of Ar has been measured to 6.0 GPa, and is shown in Fig. 12.7[54,55]. Theoretical calculation of the melting curve based on lattice dynamics and liquid perturbation theory is in very good agreement with experiment[55]. These calculations also illustrate the accuracy of the Lindemann and Ashcroft-Lekner scaling laws (see chapter 3) for melting in simple molecular systems like Ar.

12.5 Krypton

Solid Kr is fcc at RP. DAC measurements at RT to 55 GPa indicate no change of phase[56,57].

Rather little theoretical work on Kr has been reported. APW calculations of the 0 K isotherm plus thermal corrections are in very good agreement with experiment up to 45 GPa[58]. The calculated metallization pressure is 316 GPa[58].

The melting curve has been measured to 1.2 GPa[59]. This curve has been calculated accurately with a hard-sphere perturbation theory for both solid and liquid states[60]. The phase diagram of Kr is shown in Fig. 12.7.

12.6 Xenon

Xe has attracted much interest because it is expected to have the lowest metallization pressure among the rare gases. Much of the DAC work on Xe to date has been focused on the metallization problem.

Solid Xe ar RP is fcc (phase I). Numerous DAC experiments have been performed up to a pressure of 230 GPa. These include optical absorption and reflectivity[61–64] and XRD measurements[64–67]. The XRD measurements indicate a phase transition to an unknown close-packed structure (II) at ca. 14 GPa, and a further transition to hcp (III) at ca. 75 GPa. No further structural transition is seen up to 172 GPa[64,67]. The optical measurements below 100 GPa indicate a closing band gap, but extrapolations to zero gap and metallization are unreliable. Two new measurements above 100 GPa show that Xe becomes metallic in the range 130–150 GPa[63,64]. The mechanism of metallization is by an indirect 5*p*-5*d* band gap closure, as shown schematically in Fig. 12.8. The optical absorption across the gap is too weak to use for detecting metallization, and instead the onset of metallic reflectivity in the infrared is measured. The carrier density at the metallic transition is so low that metallic Xe is still optically transparent.

As with Ar, shock-Hugoniot experiments in liquid Xe show a marked softening in the *P-V* curve and a strong flattening of the pyrometric temperature, due to excitation of electrons across the band gap[68,69]. This is shown in Fig. 12.9.

Many theoretical band-structure calculations have been performed on Xe, mainly in order to predict the metallization point[64,70–72]. The very

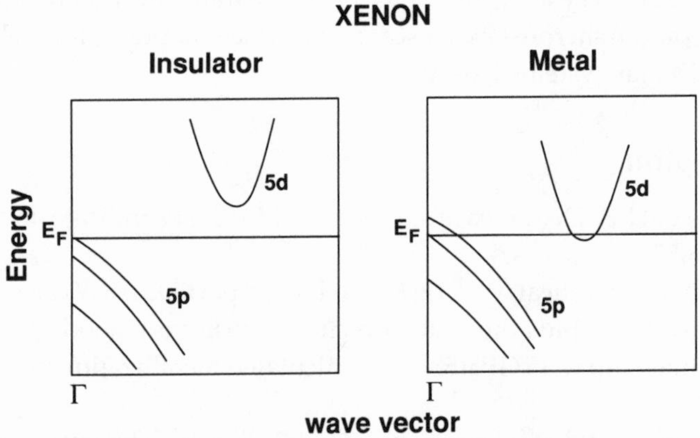

Fig. 12.8 Schematic band energies in the 100 GPa pressure range for xenon calculated by LMTO.

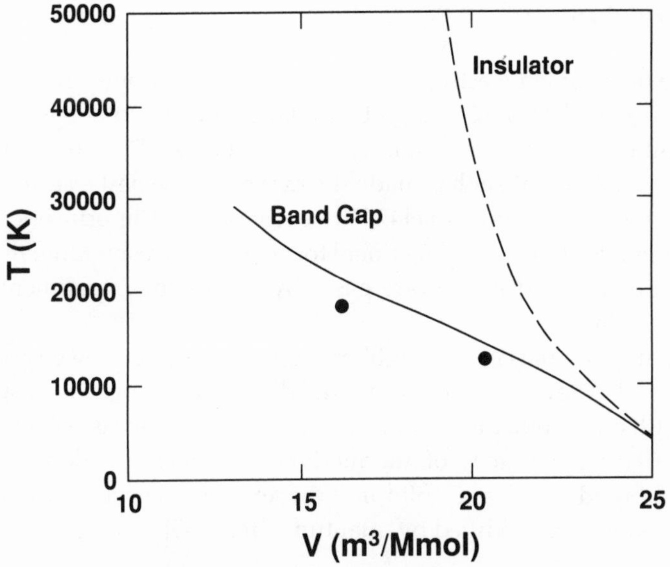

Fig. 12.9 Comparison of theoretical and experimental xenon shock temperatures. The points are experimental data. In the dashed curve, the band gap is ignored, while in the solid curve a volume-dependent band gap is included.

simple Herzfeld model predicts V(met) = 10.2 m^3/Mmol, which is remarkably accurate[74]. The various band-structure calculations have predicted pressures over the range 80–200 GPa, usually assuming an fcc lattice. The metallization pressure is sensitive to the LDA exchange-correlation potential, and to the crystal structure chosen. Calculations on hcp Xe indicate the indirect gap closure and a 2 eV interband transition which is seen in the optical experiments[63,64]. The shock-wave data are consistent with the theoretical band-gap dependence on volume[69,70].

The melting curve of Xe has been measured to 0.7 GPa[59]. A calculation with a hard-sphere perturbation theory of the solid and liquid phases together with an accurate pair potential gives good agreement with experiment[60]. The melting curve of Xe is shown in Fig. 12.7.

12.7 Radon

The crystal structure of solid Rn has not been determined[75]. Rn is intensely radioactive, with the longest-lived isotope having a half-life of only 3.8 days. Because of this, Rn is difficult to work with, and no high-pressure work has been reported. Rn melts at 202 K[76].

12.8 Discussion

Because of the filled electron shells in the rare-gas atoms, the condensed phases are insulators with large band gaps, and the total energy can be represented accurately by a sum of pair potentials. The simplicity of the pair-potential formalism has made the rare gases the best examples of the corresponding-states principle[45,77] (see chapter 3). This principle requires pair potentials of the same functional form in the corresponding materials. This can be checked in the rare gases by comparing experimental pair-potential data.

The pair potentials in the neighborhood of the attractive well are shown in Fig. 12.10. They have been compared directly in reduced form and they agree with one another to within a few percent[78]. This is reflected in the accurate correspondence of thermodynamic functions along the low-pressure liquid-vapor and solid-liquid coexistence curves, except for He and Ne, which are modified by quantum effects[45].

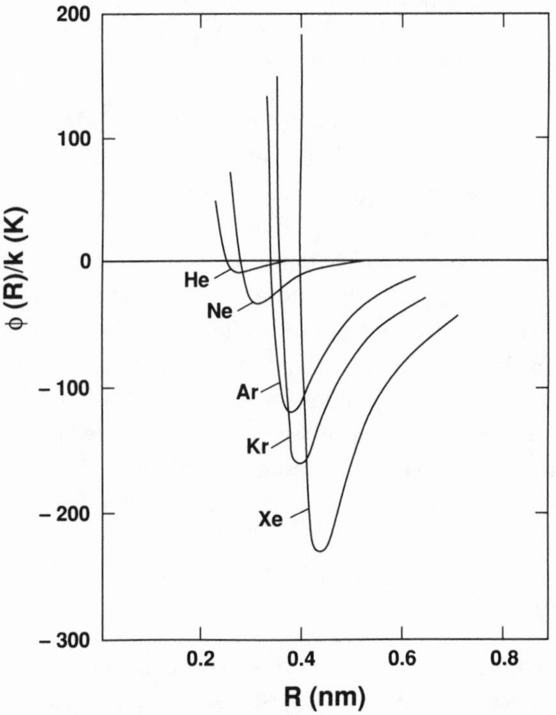

Fig. 12.10 Comparison of rare-gas pair potentials. (From R. A. Guyer, Solid State Physics **23**, 413 [1969]. Redrawn with permission.)

For very small intermolecular distances and high pressures in the solid or liquid, direct comparison can be made of isotherms, shock Hugoniots, or repulsive pair potentials. Here the correspondence is less good, because the differences in atomic structure are being probed. The breakdown of the corresponding-states principle in the rare gases can be seen clearly in the theoretical high-pressure melting curves of Kr and Xe, shown in Fig. 12.11[79]. These curves cross below 100 GPa, which they could not do if they obeyed corresponding states. The Xe melting curve is rising steeply because the metallization transition softens the effective pair potential.

A longstanding and still unsolved problem is that of the rare-gas crystal structures[80]. The problem is illustrated in Table 12.1. The predictions are the result of quasiharmonic lattice dynamics with pair potentials. In all of the solids, hcp is favored by the inverse-power multipole attractive terms. The zero-point phonons favor fcc, but this becomes decisive only in the case of He.

For He, it is likely that the very large quantum anharmonicity plays a role in stabilizing hcp. With increasing temperature, phonons stabilize the fcc structure. In the heavier rare-gas solids, it seems likely that many-body terms and crystal-field effects contribute to the static-lattice stability of fcc, but these corrections to the pair-potential energy cannot be calculated with sufficient accuracy to make the case convincing[80]. It seems that only

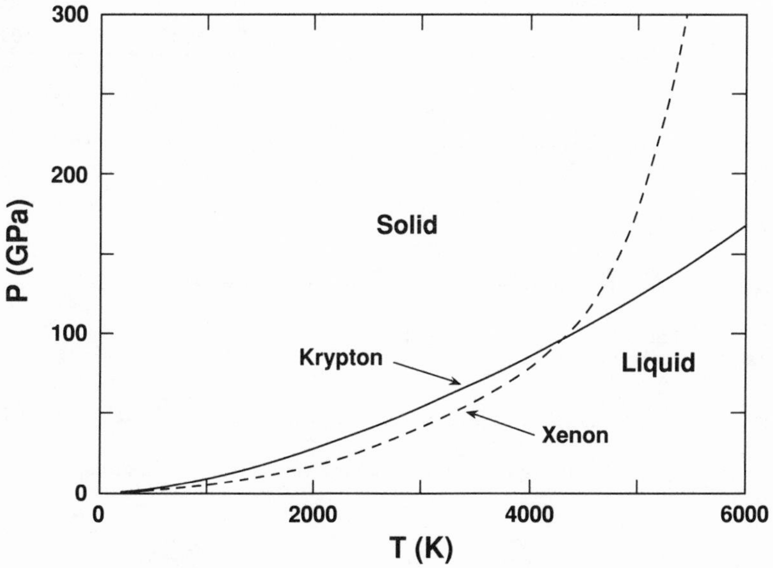

Fig. 12.11 Theoretical high-pressure melting curves of krypton and xenon.

TABLE 12.1 The Crystal-Structure Problem

Element	0 K Structure	
	Observed	Predicted
He	hcp	fcc
Ne	fcc	hcp
Ar	fcc	hcp
Kr	fcc	hcp
Xe	fcc	hcp

super-accurate band-structure calculations which intrinsically include all many-body contributions to the total energy will be able to settle this problem. A summary of rare-gas solid structures at high pressure is shown in Fig. 12.12.

At pressures above 100 GPa, band-structure contributions to the total energy invalidate the simple insulating pair-potential concept. These effects stabilize the hcp phase, in which metallization occurs. The metallization pressure of the rare-gas solids does not follow a monotonic order, which can be explained in terms of the character of the bands that cross at metallization.

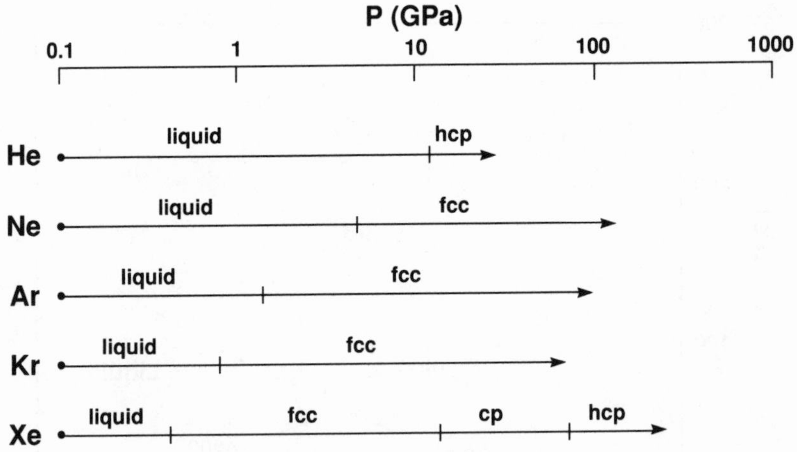

Fig. 12.12 Summary of RT structural data for the rare gases.

For Further Reading

J. Wilks, *The Properties of Liquid and Solid Helium* (Oxford University Press, Oxford, 1967).

K. H. Bennemann and J. B. Ketterson, eds., *The Physics of Liquid and Solid Helium* (Wiley, New York, 1976) 2 vols.

M. L. Klein and J. A. Venables, eds., *Rare Gas Solids* (Academic, London, 1976) 2 vols.

G. L. Pollack, "The Solid State of Rare Gases," Rev. Mod. Phys. **36**, 748 (1964).

References

1. J. Wilks, *The Properties of Liquid and Solid Helium* (Oxford University Press, Oxford, 1967).
2. A. J. Leggett, Rev. Mod. Phys. **47**, 331 (1975).
3. J. C. Wheatley, Rev. Mod. Phys. **47**, 415 (1975).
4. Ref. 1, chaps 16, 22.
5. Ref. 1, chaps. 16, 21.
6. Ref. 1, chaps. 5, 17.
7. Ref. 1, chap. 17.
8. O. V. Lounasmaa, Contemp. Phys. **15**, 353 (1974).
9. Ref. 1, chap. 11.
10. D. M. Ceperley and E. L. Pollock, Phys. Rev. Lett. **56**, 351 (1986).
11. M. H. Kalos, M. A. Lee, P. A. Whitlock, and G. V. Chester, Phys. Rev. B **24**, 115 (1981).
12. J.-P. Hansen, Phys. Lett. **30A**, 214 (1969).
13. Ref. 1, chaps. 21, 22.
14. Ref. 1, p. 475.
15. Ref. 1, p. 477.
16. J. P. Franck, Phys. Rev. B **22**, 4315 (1980).
17. J. P. Franck and W. B. Daniels, Phys. Rev. B **24**, 2456 (1981).
18. M. G. Ryschkewitsch, J. P. Franck, B. J. Duch, and W. B. Daniels, Phys. Rev. B **26**, 5276 (1982).
19. A. Driessen, Thesis, University of Amsterdam, 1982.
20. E. R. Grilly and R. L. Mills, Ann. Phys. (N.Y.) **8**, 1 (1959).
21. P. Loubeyre, J. M. Besson, J. P. Pinceaux, and J.-P. Hansen, Phys. Rev. Lett. **49**, 1172 (1982).
22. J. M. Besson, R. le Toullec, P. Loubeyre, J. P. Pinceaux, and J. P. Hansen, in *High Pressure in Science and Technology*, C. Homan, R. K. MacCrone, and E. Whalley, eds. (North-Holland, New York, 1984) part II, p. 13.
23. W. L. Vos, M. G. E. van Hinsberg, and J. A. Schouten, Phys. Rev. B **42**, 6106 (1990).
24. H. K. Mao, R. J. Hemley, Y. Wu, A. P. Jephcoat, L. W. Finger, C. S. Zha, and W. A. Bassett, Phys. Rev. Lett. **60**, 2649 (1988).
25. B. L. Holian, W. D. Gwinn, A. C. Luntz, and B. J. Alder, J. Chem. Phys. **59**, 5444 (1973).
26. P. Loubeyre, Phys. Rev. Lett. **58**, 1857 (1987).

27. D. Levesque, J.-J. Weis, and M. L. Klein, Phys. Rev. Lett. **51**, 670 (1983).
28. P. Loubeyre, Physica **139&140B**, 224 (1986).
29. M. Ross and D. A. Young, Phys. Lett. A **118**, 463 (1986).
30. D. A. Young, A. K. McMahan, and M. Ross, Phys. Rev. B **24**, 5119 (1981).
31. J. Meyer-ter-Vehn and W. Zittel, Phys. Rev. B **37**, 8674 (1988).
32. R. L. Mills, D. H. Liebenberg, and J. C. Bronson, Phys. Rev. B **21**, 5137 (1980).
33. W. J. Nellis, N. C. Holmes, A. C. Mitchell, R. J. Trainor, G. K. Governo, M. Ross, and D. A. Young, Phys. Rev. Lett. **53**, 1248 (1984).
34. P. Loubeyre and J.-P. Hansen, Phys. Rev. B **31**, 634 (1985).
35. P. Loubeyre and J.-P. Hansen, Phys. Lett. **80A**, 181 (1980).
36. J.-L. Barrat, P. Loubeyre, and M. L. Klein, J. Chem. Phys. **90**, 5644 (1989).
37. P. B. Foreman, P. K. Rol, and K. P. Coffin, J. Chem. Phys. **61**, 1658 (1974).
38. D. M. Ceperley and H. Partridge, J. Chem. Phys. **84**, 820 (1986).
39. R. A. Aziz, F. R. W. McCourt, and C. C. K. Wong, Mol. Phys. **61**, 1487 (1987).
40. C. A. Swenson, in *Rare Gas Solids*, M. L. Klein and J. A. Venables, eds. (Academic, London, 1977) vol. 2, chap. 13.
41. L. W. Finger, R. M. Hazen, G. Zou, H. K. Mao, and P. M. Bell, Appl. Phys. Lett. **39**, 892 (1981).
42. R. J. Hemley, C. S. Zha, A. P. Jephcoat, H. K. Mao, L. W. Finger, and D. E. Cox, Phys. Rev B **39**, 11820 (1989).
43. J. C. Boettger, Phys. Rev. B **33**, 6788 (1986).
44. J. Hama, Phys. Lett. **105A**, 303 (1984).
45. R. K. Crawford, in Ref. 40, chap. 11.
46. W. L. Vos, J. A. Schouten, D. A. Young, and M. Ross, preprint.
47. Ref. 40, vols. 1 and 2.
48. M. Ross, H. K. Mao, P. M. Bell, and J. A. Xu, J. Chem. Phys. **85**, 1028 (1986).
49. W. J. Nellis and A. C. Mitchell, J. Chem. Phys. **73**, 6137 (1980).
50. S. P. Marsh, ed., *LASL Shock Hugoniot Data* (University of California Press, Berkeley, Los Angeles, London, 1980) pp. 16–18.
51. F. V. Grigor´ev, S. B. Kormer, O. L. Mikhailova, M. A. Mochalov, and V. D. Urlin, Zh. Eksp. Teor. Fiz. **88**, 1271 (1985) [Sov. Phys. JETP **61**, 751 (1985)].
52. A. K. McMahan, Phys. Rev. B **33**, 5344 (1986).
53. M. Ross, J. Chem. Phys. **73**, 4445 (1980).
54. R. G. Crafton, Phys. Lett. **36A**, 121 (1971).
55. C.-S. Zha, R. Boehler, D. A. Young, and M. Ross, J. Chem. Phys. **85**, 1034 (1986).
56. I. V. Aleksandrov, A. N. Zisman, and S. M. Stishov, Zh. Eksp. Teor. Fiz. **92**, 657 (1987) [Sov. Phys. JETP **65**, 371 (1987)].
57. A. Polian, J. M. Besson, M. Grimsditch, and W. A. Grosshans, Phys. Rev. B **39**, 1332 (1989).
58. J. Hama and K. Suito, Phys. Lett. A **140**, 117 (1989).
59. P. H. Lahr and W. G. Eversole, J. Chem. Eng. Data **7**, 42 (1962).
60. J. H. Kim, T. Ree, and F. H. Ree, J. Chem. Phys. **91**, 3133 (1989).
61. K. Syassen, Phys. Rev. B **25**, 6548 (1982).
62. K. Asaumi, T. Mori, and Y. Kondo, Phys. Rev. Lett. **49**, 837 (1982).
63. K. Goettel, J. H. Eggert, I. F. Silvera, and W. C. Moss, Phys. Rev. Lett. **62**, 665 (1989).

64. R. Reichlin, K. E. Brister, A. K. McMahan, M. Ross, S. Martin, Y. K. Vohra, and A. L. Ruoff, Phys. Rev. Lett. **62**, 669 (1989).
65. K. Asaumi, Phys. Rev. B **29**, 7026 (1984).
66. A. N. Zisman, I. V. Aleksandrov, and S. M. Stishov, Phys. Rev. B **32**, 484 (1985).
67. A. P. Jephcoat, H. K. Mao, L. W. Finger, D. E. Cox, R. J. Hemley, and C.-S. Zha, Phys. Rev. Lett. **59**, 2670 (1987).
68. W. J. Nellis, M. van Thiel, and A. C. Mitchell, Phys. Rev. Lett. **48**, 816 (1982).
69. H. B. Radousky and M. Ross, Phys. Lett. A **129**, 43 (1988).
70. M. Ross and A. K. McMahan, Phys. Rev. B **21**, 1658 (1980).
71. A. K. Ray, S. B. Trickey, R. S. Weidman, and A. B. Kunz, Phys. Rev. Lett. **45**, 933 (1980).
72. J. Hama and S. Matsui, Solid State Comm. **37**, 889 (1981).
73. A. K. Ray, S. B. Trickey, and A. B. Kunz, Solid State Comm. **41**, 351 (1982).
74. M. Ross, Rep. Prog. Phys. **48**, 1 (1985).
75. J. Donohue, *The Structures of the Elements* (Wiley, New York, 1974) chap. 2.
76. R. C. Weast and M. J. Astle, eds., *CRC Handbook of Chemistry and Physics*, 63d ed. (CRC Press, Boca Raton, 1982) p. B-34.
77. G. L. Pollack, Rev. Mod. Phys. 3**6**, 748 (1964).
78. J. A. Barker, in Ref. 40, vol. 1, chap. 4.
79. F. H. Ree, personal communication.
80. K. F. Niebel and J. A. Venables, in Ref. 40, vol. 1, chap. 9.

CHAPTER 13
The Transition Metals

13.1 Introduction

The transition metals represent the filling of the atomic *d*-electron shell. I have included 29 elements in the group, representing the 3*d*, 4*d*, and 5*d* shells. Although the last column, containing Zn, Cd, and Hg, is usually considered a separate group, it is included here for convenience. The elements La and Ac are included in the lanthanides and actinides, respectively.

Because the transition metals and their alloys typically have high melting temperatures and hardness, their economic importance is immense. The need to control the properties of alloys has stimulated theoretical work on the phase relations of the pure transition metals. This effort has recently achieved success in the theoretical explanation of the RPT crystal structures of the transition metals using one-electron band-structure theory. Also, there is increasing experimental and theoretical work on liquid transition metals.

There has been rather little experimental work on phase transitions in transition metals because these elements are relatively incompressible, yielding phase changes only at very high pressures beyond the experimental range. Also, the melting temperatures are high, making the melting curves difficult to measure. Shock waves are at present the most effective means of studying the high-pressure phase diagrams of the transition metals.

In this chapter I survey the experimental and theoretical work on the transition metals one column (3 elements) at a time, and then summarize the group as a whole in the final discussion.

13.2 The Scandium Group

13.2.1 Scandium

At RTP, Sc I is hcp. Compression at RT indicates a sluggish transition near 20 GPa to Sc III, which is st(4)[1–3]. This structure may be identical with that of β-Np, but the precise space group is uncertain[1]. Compression to 45 GPa shows no further phase transitions[1]. Shock-compression experiments indicate a change in slope of the u_s-u_p curve at ca. 37 GPa[4], which may be correlated with the I-III transition.

At RP and 1608 K, the hcp phase transforms to bcc Sc II[5]. Sc II melts at 1814 K. These transitions have not been studied at high pressure.

13.2.2 Yttrium

At RTP, Y is hcp. Compression at RT leads to three transitions: hcp→hex(9) Sm-type at ca. 10 GPa, hex(9)→hex(4) dhcp at ca. 26 GPa, and dhcp→fcc at ca. 39 GPa[1,6]. It is possible that above 40 GPa the fcc lattice undergoes a distortion to an hex(6) symmetry[1,7]. These phases are typical of the lanthanide elements. No further transitions are seen up to 47 GPa. There is an anomaly in the shock Hugoniot near 47 GPa which may be related to these phase transitions[4].

A transition from hcp to a new phase, probably bcc, occurs at RP and 1752 K[5]. Y melts at 1795 K. Neither transition has been studied at high pressure.

13.2.3 Theoretical

The scandium group, with electronic configuration s^2d^1, is expected to behave like the trivalent lanthanides. Canonical-band[8] and LMTO[9] theory correctly predict the RTP hcp structures of Sc and Y, and the lanthanide sequence in Y. AIP calculations on Y also predict not only that hcp is the stable phase but that the c/a ratio is 1.57, in agreement with observation[10]. However, these calculations predict an unrealistically low transition pressure to fcc. The lanthanide sequence in Y proves that this sequence has nothing to do with 4*f* electrons but is a phenomenon of the *s-d* electron transfer. The transitions in Y are well correlated theoretically with the progress of the *s-d* transfer and the increase of the *d*-electron number n_d from 1.4 (hcp) to 2.7 (fcc)[9].

The *s-d* electron transfer, as in the Group I and II elements, is accompanied by an increase in compressibility[11] and a decrease in the Grüneisen parameter, which account for the shock anomalies found in Sc and Y.

13.3 The Titanium Group

13.3.1 Titanium

At RTP, α-Ti is hcp. At ca. 2 GPa and RT, Ti transforms to the ω phase which is hex(3). This structure may be regarded as a hexagonal distortion of bcc[12,13]. The α–ω phase line has been found to have a negative slope[14–17]. This transition shows a large hysteresis and the equilibrium phase boundary is not accurately known. Further compression at RT to 87 GPa shows only the hex(3) phase[18].

At RP and 1155 K, the hcp α phase transforms to bcc β, which is denser. The α–β phase boundary has been determined in high-temperature static experiments[14,19]. The α, β, and ω phases come together in a triple point at ca. 940 K and 9.0 GPa. The ω–β phase boundary has been determined to 15 GPa[14]. No other phases have been found.

Shock-wave experiments on Ti show a discontinuity in the u_s-u_p curve, which has been interpreted as the α-ω or ω-β transition[20–22]. However, this feature may be an artifact of the elastic wave precursor in the shock[23].

Ti melts at 1943 K. The melting curve has not been determined, but its slope is known from isobaric-heating measurements of ΔH and ΔV across the melting transition[24]. The phase diagram of Ti is shown in Fig. 13.1.

13.3.2 Zirconium

At RTP, α-Zr is hcp. At ca. 2 GPa and RT, Zr transforms to the hex(3) ω phase, as in Ti. The α–ω phase line has a negative slope[15,16,19,25], but hysteresis

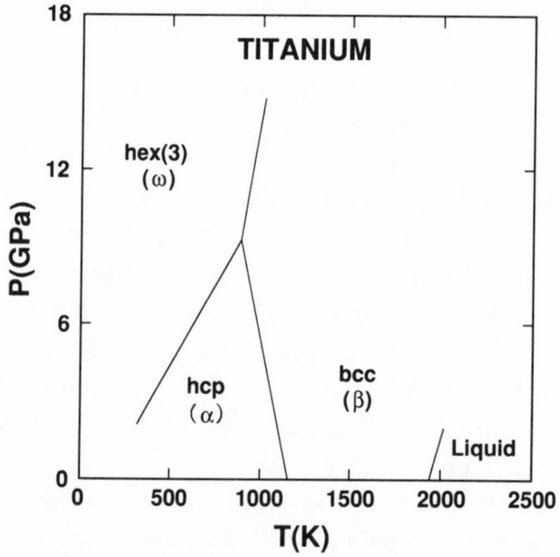

Fig. 13.1 The phase diagram of titanium.

prevents an accurate determination of the equilibrium transition pressure. DAC experiments at RT find an ω-to-bcc transition at 30 GPa[26].

At RP and 1136 K, the α hcp phase transforms to the β bcc phase. The α–β phase boundary has been determined in high-temperature static experiments[15,19]. The α–β–ω triple point is found to be at 975 K and 6.7 GPa[25]. The ω–β phase boundary has been determined to 7.5 GPa[19]. The β phase is probably identical with the high-pressure bcc phase found at RT, so the ω-β phase boundary must turn backward toward the $T = 0$ K axis at high pressure.

As in Ti, shock-wave experiments on Zr show a discontinuity in the u_s-u_p curve, which has been interpreted as a phase transition[20,27].

Zr melts at 2128 K. The melting curve has not been determined, but its slope is known from isobaric-heating measurements[24]. The phase diagram of Zr is shown in Fig. 13.2.

13.3.3 Hafnium

At RTP, α-Hf is hcp. Compression of Hf at RT to 38 GPa yields the ω phase, and further compression to 71 GPa yields the bcc phase[18]. No further transitions are found up to 252 GPa[18].

At RP and 2030 K, the α hcp phase transforms to the β(bcc) phase, which is denser, as in Ti and Zr[5,12]. The α-β phase boundary has not been determined.

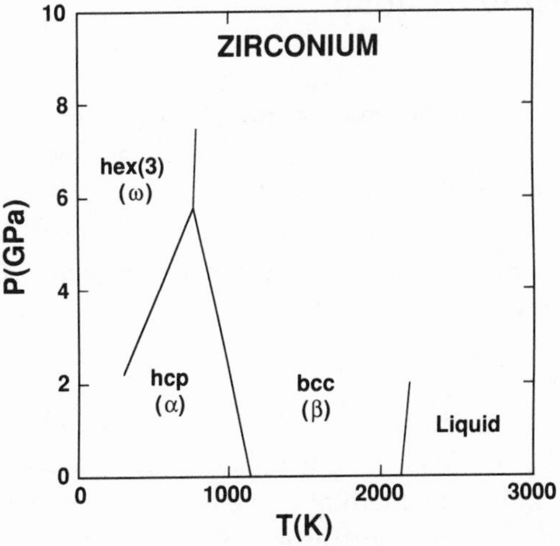

Fig. 13.2 The phase diagram of zirconium.

Shock-wave experiments on Hf show a discontinuity in the u_s-u_p curve, which has been interpreted as a phase transition[20].

Hf melts at 2504 K. The melting curve has not been determined.

13.3.4 Theoretical

LMTO calculations for Ti which include the hcp, bcc, ω, and fcc structures predict that the ω phase will be stable over the pressure range 0–30 GPa[28]. It is possible, considering the α-ω trajectory of Fig. 13.1, that ω is in fact the equilibrium phase of Ti at 0 K, and that the LMTO prediction is correct. For Zr, LMTO calculations predict an α-ω transition at 5 GPa and an ω-bcc transition at 11 GPa[18]. LMTO calculations for Hf predict an hcp-bcc transition at 51 GPa[29]. The absence of predicted ω stability in Hf may be due to the inadequacy of the atomic-sphere approximation for the ω phase. LMTO and other linearized band-structure calculations without the ω phase predict hcp stability for Ti, Zr, and Hf[9,30]. The calculations are in semiquantitative agreement with experiment. These results, together with the appearance of ω at high pressure and at RP in alloys such as TiV and ZrNb, and the similarity of the β and ω structures, all suggest that the phase diagrams of the titanium group exhibit the *s-d* transfer phenomenon[13]. Increasing the *d*-electron number either by pressure or by alloying with *d*-rich elements, drives the structure toward the bcc phase characteristic of the next group of elements to the right[13]. The specific form of the ω structure as a hexagonal distortion of the bcc phase may be related to the details of the Fermi surface[31,32].

13.4 The Vanadium Group

13.4.1 Vanadium

At RTP, V is bcc. XRD measurements with NaCl as a pressure scale showed no phase transition up to 10 GPa[33]. Shock-Hugoniot measurements to 340 GPa show no sign of a phase transition[34].

V melts at 2183 K. The melting curve has not been measured directly, but the slope is known from isobaric-heating experiments[35]. The phase diagram of V is shown in Fig. 13.3.

13.4.2 Niobium

At RTP, Nb is bcc. Static compression to 25 GPa[33,36] and shock-wave experiments[37] up to 170 GPa show no phase transitions.

Nb melts at 2742 K. The melting curve has not been measured but the slope is known from isobaric-heating experiments[38]. The phase diagram of Nb is shown in Fig. 13.3.

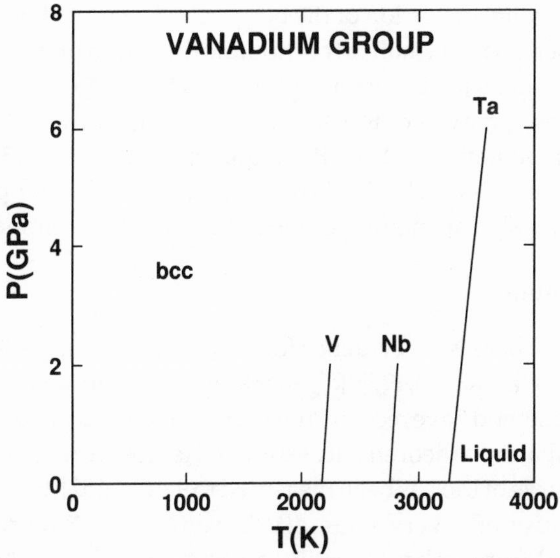

Fig. 13.3 The melting curves of the vanadium group.

13.4.3 Tantalum

At RTP, Ta is bcc. Static compression at RT to 77 GPa shows no phase transition[33,39]. There is no obvious anomaly corresponding to a solid-solid transition on the shock Hugoniot[37,40].

The melting curve of Ta has been measured to 6.0 GPa[41]. This curve is in good agreement with the initial melting slope obtained from isobaric-heating data[42]. The optical-analyzer technique has been used to locate the melting point on the shock Hugoniot of Ta at 295 GPa[43]. The melting temperature theoretically computed at this point is approximately 10000 K. The phase diagram of Ta is shown in Fig. 13.3.

13.4.4 Theoretical

Linearized band-structure[9,30] and more approximate GPT calculations[44] predict the observed bcc structure of the vanadium group and show that bcc is strongly preferred over fcc or hcp. No calculations are available for other possible crystal structures of this group.

13.5 The Chromium Group

13.5.1 Chromium

At RPT, Cr is bcc. This structure is modified very slightly by two first-order magnetic phase transitions[45]. From 0 K to 123 K Cr is antiferromagnetic with a small tetragonal distortion of the bcc lattice. From 123 K to 311 K, Cr is antiferromagnetic with a small orthorhombic distortion of the bcc lattice. These phase transitions have been measured to 0.8 GPa[46,47]. The lattice distortions in these phases are too small to be detected by XRD. Above 311 K, Cr is paramagnetic and bcc. RT compression to 10 GPa[33,48] and shock compression to 140 GPa[37] show no evidence of further phases.

Cr melts at 2136 K. The melting curve has not been determined.

13.5.2 Molybdenum

Mo is bcc at RTP. There is no evidence for phase changes up to 280 GPa in DAC experiments at RT[33,49,50]. Shock-wave experiments with the optical-analyzer method have recently revealed a solid-solid phase transition in Mo at ca. 210 GPa and a theoretically estimated temperature of 4100 K[51].

The melting curve of Mo has been measured to 9.0 GPa[52], but these data are suspect because of a very large dP/dT value[53]. A more realistic dP/dT is obtained from isobaric-heating measurements[42]. The optical-analyzer technique has also been used to find the melting point on the shock Hugoniot at 390 GPa and an estimated temperature of 10000 K[51]. The phase diagram of Mo is shown in Fig. 13.4.

13.5.3 Tungsten

At RTP, W is bcc. Static compression to 364 GPa shows no phase change[54]. Optical-analyzer experiments show a discontinuity in the sound speed along the Hugoniot at ca. 440 GPa which might correspond to a solid-solid phase transition[55].

The melting curve of W has been determined to 5.0 GPa[56]. Isobaric-heating data are in rough accord with this melting slope[57]. The optical-analyzer experiments suggest that melting occurs at ca. 490 GPa on the Hugoniot[55]. The phase diagram of W is shown in Fig. 13.4.

13.5.4 Theoretical

According to a number of different band-structure theories[8,9,30,58,59], the chromium group is predicted to have the bcc structure. LAPW calculations of magnetic bcc Cr using the local-spin-density approximation (LSDA) are in poor agreement with the measured bulk properties of Cr, which probably indicates the failure of the LSDA[60]. LMTO calculations on the fcc, hcp, and bcc phases of Mo show that the bcc phase is destabilized by pressure and transitions to hcp at 320 GPa and then to fcc at 470 GPa are expected[51]. It is possible that the solid-solid phase transition found in

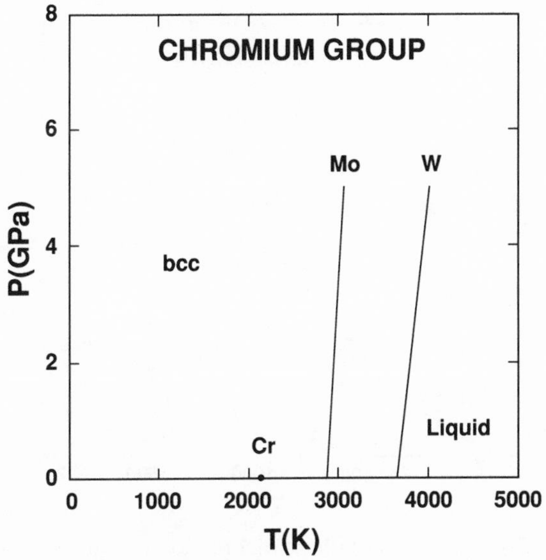

Fig. 13.4 The melting curves of the chromium group.

shocked Mo is the bcc-hcp transition. LMTO calculations on Cr and W also predict a bcc-hcp transition, but at pressures well above the Mo value[61]. It is possible that the predicted transition in W is seen in the optical-analyzer results. The LMTO results again demonstrate the *s–d* transfer, in which pressure increases *d*-electron number and drives structures toward those of higher *d*-number elements. Thus Mo under pressure looks like Tc, which is hcp.

13.6 The Manganese Group

13.6.1 Manganese

Mn exhibits a complex phase diagram with four phases observed at RP. From 0 K to ca. 1000 K, α-Mn is bcc(58), with 4 different types of atomic sites[12]. RT compression of α-Mn to 42 GPa shows no phase transition[62]. Between 1000 K and 1370 K, β-Mn is sc(20), with 2 types of atomic sites[12]. The α- and β-Mn structures are unique among the elements. Between 1370 K and 1410 K, γ-Mn is fcc, and from 1410 K to the melting point at 1517 K, δ-Mn is bcc. The three solid-solid phase transitions have been measured to 4.0 GPa[63].

The melting curve of Mn has been measured to 4.0 GPa[63]. The phase diagram of Mn is shown in Fig. 13.5.

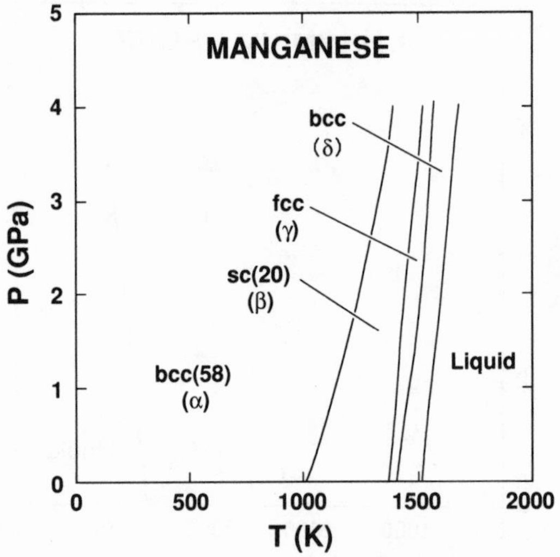

Fig. 13.5 The phase diagram of manganese.

13.6.2 Technetium

Tc is an unstable element with a half-life of 4×10^6 yr. Solid Tc at RTP is hcp. Tc has been compressed to 10 GPa with no indication of any phase change[64].

Tc melts at 2428 K. The melting curve has not been determined.

13.6.3 Rhenium

Solid Re is hcp at RTP. Re is a preferred gasket material in very-high-pressure DAC configurations, and it has been compressed at RT to 251 GPa with no change of phase[65,66].

The melting curve of Re has been measured to 8.0 GPa[67]. The phase diagram of Re is shown in Fig. 13.6.

13.6.4 Theoretical

The complex structures observed in α- and β-Mn are likely to be the effect of its antiferromagnetism, but this has not yet been confirmed theoretically. The other two elements, Tc and Re, follow the pattern of structures seen across the transition-metal series. Linearized band-structure predictions agree with the observed hcp structures for Tc and Re[9,30]. Linear augmented Slater-type orbitals calculations on bcc and hcp Re show that hcp will remain the stable phase over the DAC pressure range[68].

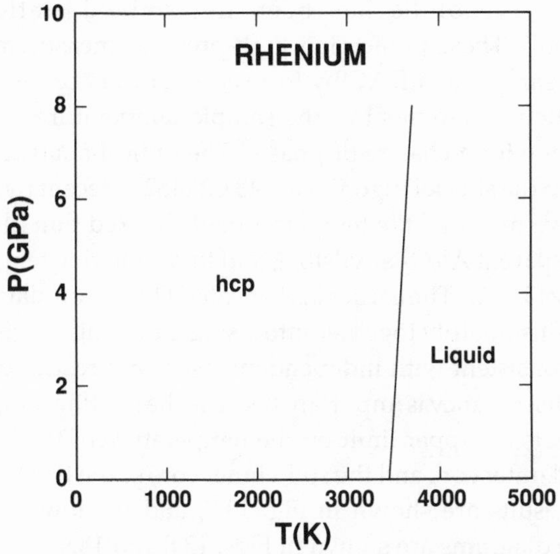

Fig. 13.6 The phase diagram of rhenium.

13.7 The Iron Group

13.7.1 Iron

Iron is not only important technologically but is also the main constituent of the Earth's core. This has stimulated a continuing research effort, both experimental and theoretical, on the iron phase diagram.

At RTP, α-Fe is bcc and ferromagnetic. RT compression to ca. 13 GPa leads to the ε hcp phase[69–72]. This phase transition is sluggish and the equilibrium phase boundary is sensitive to nonhydrostatic stresses. The ε phase is nonmagnetic[73]. Further RT compression to 304 GPa shows no phase transition[74,75]. At RP and 1043 K, α-Fe passes through the Curie point, and at 1173 K, it transforms to paramagnetic γ fcc. The α-γ, α-ε, and γ-ε phase boundaries have been studied intensively[69–72,76–81], and the α-γ-ε triple point occurs at ca. 757 K and 10.4 GPa[76].

The γ-ε phase boundary has been measured to 50 GPa by static techniques[72,77,80,81]. Optical-analyzer measurements on Fe show a discontinuity in sound speed at 200 GPa, which may be a continuation of the γ-ε transition[82]. However, linear extrapolation of the experimental γ-ε phase boundary suggests that it intersects the melting curve well below 200 GPa, and that the shock discontinuity is due to the appearance of a new phase[81,83]. At RP and 1660 K, γ-Fe transforms to δ bcc. This phase disappears in a triple point on the melting curve at 1991 K and 5.3 GPa[84].

The melting curve of Fe has been determined statically to 120 GPa[72,81,8–86]. These pioneering high-pressure measurements involve heating the sample in a DAC by focused laser light or by electrical current, pyrometric measurement of the sample temperature, and some diagnostic technique for a change of phase. The optical-analyzer results indicate melting on the shock Hugoniot at 243 GPa[82]. Recent pyrometric measurements have recorded the temperature of shocked thin films of Fe deposited on transparent Al_2O_3 anvils[87], and find a melting temperature of 6700 ± 100 K at 243 GPa. The earlier laser-heated DAC static data and the shock-wave data fit smoothly together into a single melting curve[86], but this curve is not consistent with independent laser- and resistive-heating data[72,81]. This discrepancy is important because the melting temperature of Fe at 330 GPa sets an upper limit on the temperature of the boundary between the liquid outer core and the solid inner core of the Earth[86]. The optical-analyzer results are shown in Fig. 13.7, and the low- and high-pressure Fe phase diagrams are shown in Figs. 13.8 and 13.9.

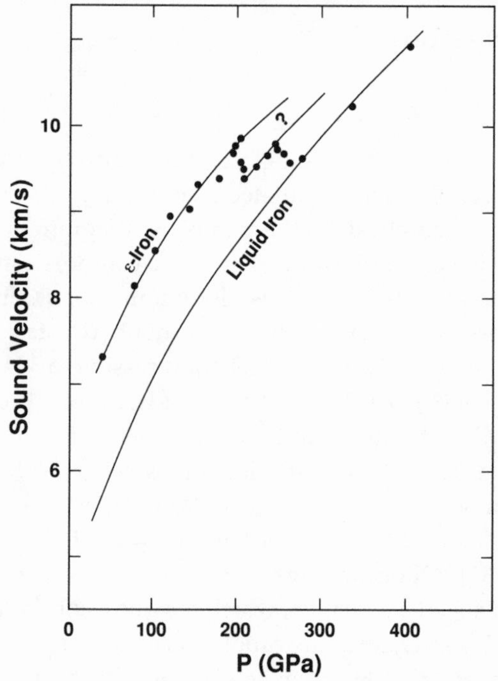

Fig. 13.7 Rarefaction velocities of iron on the Hugoniot as a function of pressure. (From Brown and McQueen[82]. © American Geophysical Union. Redrawn with permission.)

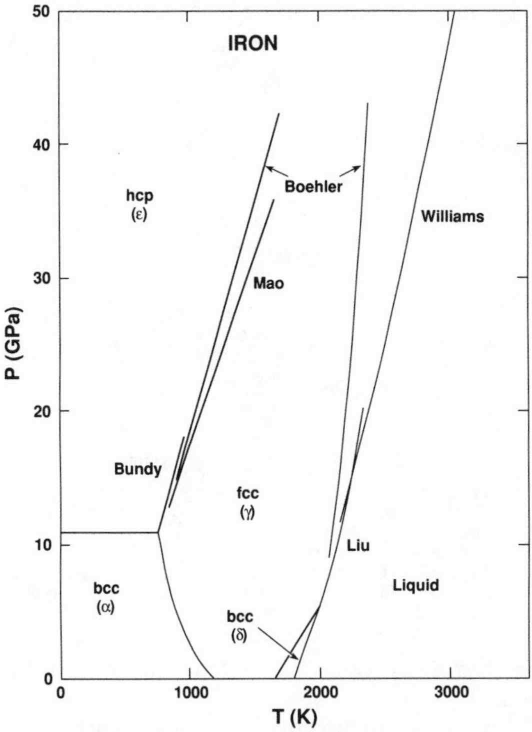

Fig. 13.8 The low-pressure phase diagram of iron.

13.7.2 Ruthenium

At RTP, Ru is hcp. Compression to 25 GPa at RT shows no change of phase[88].

Ru melts at 2607 K. Melting curve measurements have not been reported.

13.7.3 Osmium

At RTP Os is hcp. High-pressure work on Os has not been reported.

Os melts at 3306 K. The melting curve has not been reported.

13.7.4 Theoretical

Fe has been studied theoretically in depth, but a clear understanding of the phase diagram does not yet exist. Nonmagnetic LMTO calculations predict hcp to be the stable phase at RTP[9]. Attempts to predict the correct magnetic ground state of Fe with spin-polarized band-structure calculations have failed, giving fcc as the stable phase[89,90]. In addition, there are large errors in the predicted volume and bulk modulus of Fe[91]. The

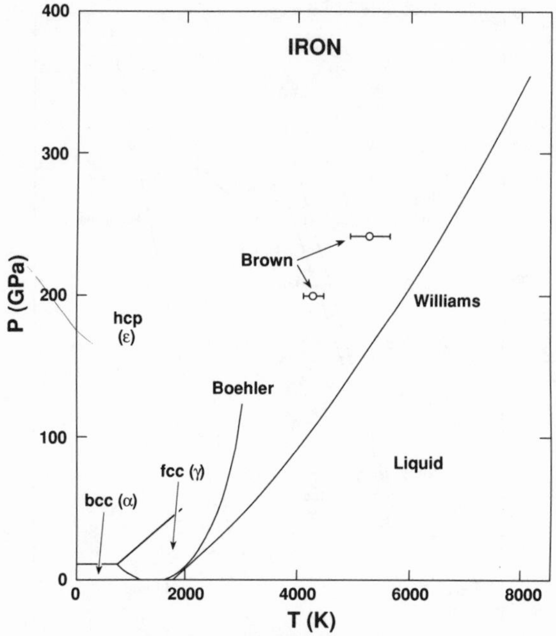

Fig. 13.9 The high-pressure phase diagram of iron.

problem may lie in the local-spin-density approximation to the exchange-correlation potential[89], since nonlocal gradient corrections to the LSDA correctly predict the ferromagnetic bcc ground state[92].

On a more approximate level, there have been semiempirical models of the overall Fe phase diagram. The most complete model suggests that the sequence of phase changes is driven by magnetism[93], but another theory ascribes the phase changes to electronic and lattice-dynamics effects[94].

The melting curve of Fe is of great interest in connection with the physics of the Earth's core, and numerous widely divergent predictions of this curve up to core pressures (330 GPa) have been made[95–98]. The melting models are all semiempirical because the interatomic potential in solid and liquid Fe is poorly understood.

For Ru and Os, linearized band-structure calculations correctly predict the hcp structure[9,30].

13.8 The Cobalt Group

13.8.1 Cobalt

At RTP, Co is hcp. There is a martensitic transition to fcc at RP and 695 K[99]. The hcp-fcc phase boundary has been measured to 5.0 GPa[100].

Co melts at 1768 K. The melting curve has not been measured directly, but dP/dT may be obtained from experimental ΔS and ΔV data[99,101]. The phase diagram of Co is shown in Fig. 13.10.

13.8.2 Rhodium

At RTP, Rh is fcc. RT compression to 25 GPa shows no phase transition[102].

The melting curve of Rh has been determined to 10 GPa[103]. The phase diagram of Rh is shown in Fig. 13.11.

13.8.3 Iridium

At RTP, Ir is fcc. RT compression to 30 GPa shows no phase transition[104].

Ir melts at 2720 K. The melting curve has not been reported, but the slope may be computed from isobaric-heating experiments[35]. The phase diagram of Ir is shown in Fig. 13.11.

13.8.4 Theoretical

LMTO calculations using the local-spin-density approximation correctly predict an hcp ferromagnetic ground state for Co[105]. As in Mn and Fe, the magnetism has an important effect on the structure of Co.

Linearized band-structure calculations[9,30] also correctly predict the RTP fcc structure of Rh and Ir.

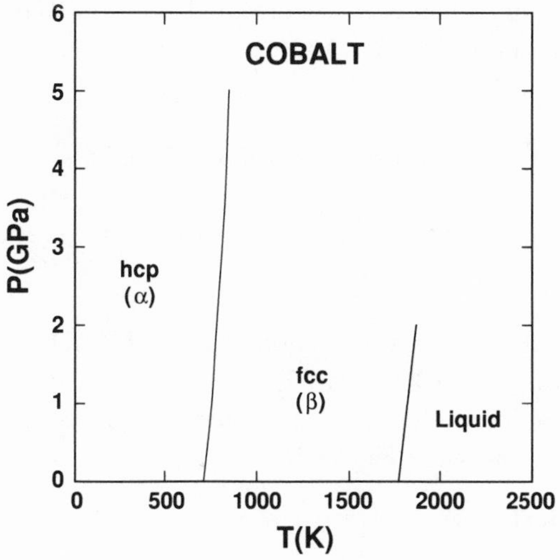

Fig. 13.10 The phase diagram of cobalt.

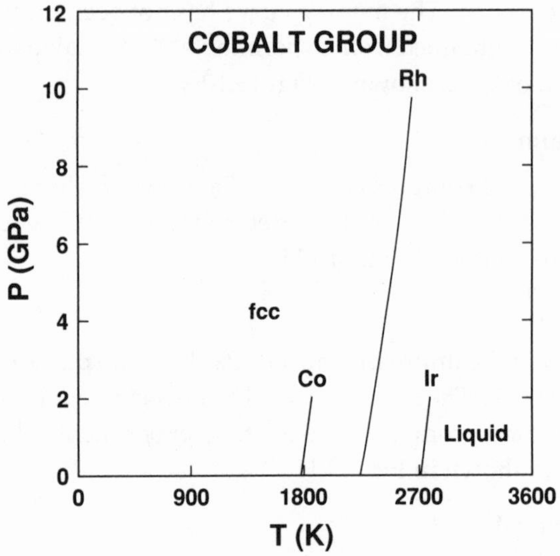

Fig. 13.11 The melting curves of the cobalt group.

13.9 The Nickel Group

13.9.1 Nickel

At RTP, Ni is fcc. The fcc structure is stable to 65 GPa[106].

The melting curve has been determined to 10 GPa[103]. The phase diagram of Ni is shown in Fig. 13.12.

13.9.2 Palladium

At RTP, Pd is fcc. RT compression of Pd to 77 GPa shows no change of phase[49].

Pd melts at 1828 K. The melting curve has not been reported.

13.9.3 Platinum

At RTP, Pt is fcc. Pt is a pressure standard for the DAC[107], and it remains in the fcc structure up to 304 GPa[75].

The Pt melting curve has been determined to 10 GPa[56,103]. The Pt phase diagram is shown in Fig. 13.12.

13.9.4 Theoretical

Linearized band-structure calculations correctly predict the fcc structure of the nickel group elements[9,30].

Because Ni has the electronic configuration $3d^8 4s^2$, it may exhibit unusual metal-to-insulator behavior at very high pressure, where the

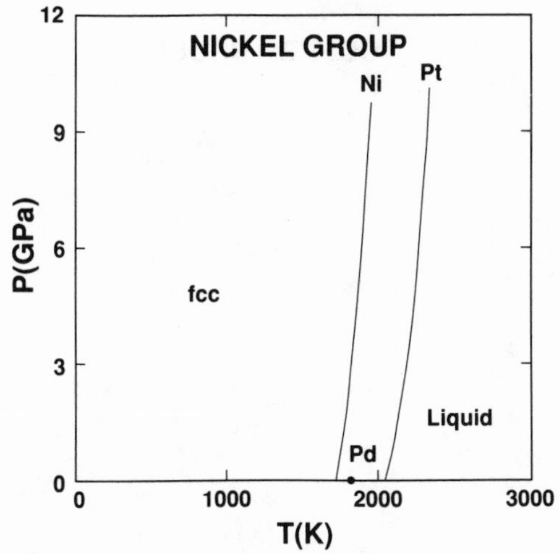

Fig. 13.12 The melting curves of the nickel group.

electronic configuration changes from $3d^8 4s^2$ to $3d^{10}$[108]. According to APW calculations[109], this occurs as the bottom of the $4s^2$ band rises above the bottom of the top of the 3*d* band at 34 TPa, leaving an insulating gap. The gap is closed again at 51 TPa as the top of the 3*d* band overlaps the bottom of the 4*f* band. The fcc structure is preferred over the hcp or bcc structures throughout. A similar metal-insulator transition does not occur in Pd or Pt because the bands are broader[110].

In the nickel group, the normal high-pressure *s-d* transfer reverses to a weak *d-s* transfer, which tends to stabilize the fcc structure under compression. LMTO calculations on Pt predict that fcc will remain stable beyond 500 GPa[107].

13.10 The Copper Group

13.10.1 Copper

At RTP, Cu is fcc. RT compression[39,111] to 188 GPa shows no phase transition.

The melting curve of Cu has been measured to 6.0 GPa[112]. The phase diagram of Cu is shown in Fig. 13.13.

13.10.2 Silver

At RTP, Ag is fcc. RT compression[49,113] to 92 GPa shows no change of phase.

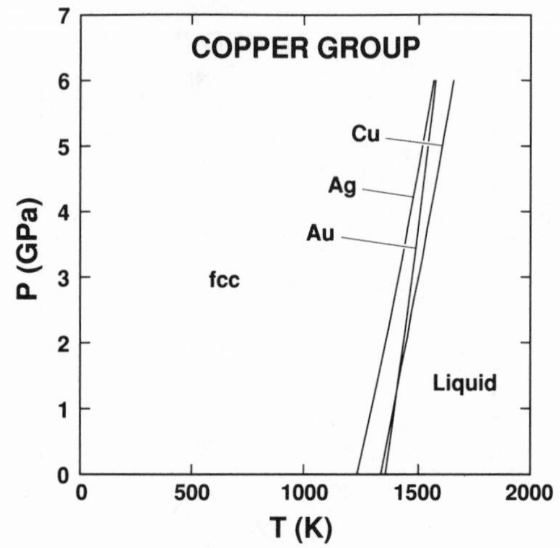

Fig. 13.13 The melting curves of the copper group.

The melting curve of Ag has been measured to 6.0 GPa[112]. The phase diagram of Ag is shown in Fig. 13.13.

13.10.3 Gold

Au is fcc at RTP. RT compression[111,114] to 185 GPa shows no change of phase. Because of its phase stability and large nuclear charge, Au is considered an excellent XRD pressure standard for the DAC both at high pressure and high temperature[114–116].

The melting curve of Au has been measured to 6.0 GPa[112]. The phase diagram of Au is shown in Fig. 13.13.

13.10.4 Theoretical

There have been numerous attempts to treat the copper group as quasi-NFE metals because of the filled *d*-band configuration s^1d^{10}. However, the presence of the *d* band has an important effect on the bonding by hybridization. LMTO calculations correctly predict fcc phase stability for Cu and Ag at RP, but bcc is incorrectly predicted by a small margin to be the stable phase for Au[9]. AIP calculations predict fcc stability for Au, but hcp stability for Ag, again by a very small margin[117]. Linearized full-potential band-structure calculations also predict fcc stability for Au[30].

As in the nickel group, the *d*-*s* transfer occurs in the copper group and stabilizes the fcc phase. GPT calculations predict the fcc structure in Cu to remain stable to beyond 1 TPa[118]. GPT also predicts a melting curve in good agreement with experiment for Cu[61,119].

13.11 The Zinc Group

13.11.1 Zinc

Zn at RTP has a distorted hcp structure, with the c/a ratio considerably greater than ideal[12]. This ratio decreases smoothly toward the ideal value under compression, but no phase transition is observed up to 30 GPa[120].

The melting curve has been determined to 6.0 GPa[121]. The phase diagram of Zn is shown in Fig. 13.14.

13.11.2 Cadmium

Cd is hcp, with a nonideal c/a ratio. Compression to 39 GPa shows a smooth decrease in this ratio, but no phase transition[120].

The melting curve has been determined to 4.0 GPa[122]. The phase diagram of Cd is shown in Fig. 13.14.

13.11.3 Mercury

At low temperatures β-Hg has a ct(2) structure. At 78 K there is a transformation to α rh(1). Both α and β are simple distortions of close-packed structures[12]. The α-β phase boundary has been determined to ca. 20 GPa[123,124]. This boundary has a maximum temperature at ca. 7.5 GPa, and above this pressure, the α-β temperature decreases. The α-β transition is sluggish in this region and the phase boundary is poorly determined. At RT and pressures above 30 GPa, the α phase transforms to δ hcp[124]. The α-δ transition also has a very large hysteresis. The α-β-δ triple point is estimated at 27 GPa and 180 K. A fourth phase named γ has been found

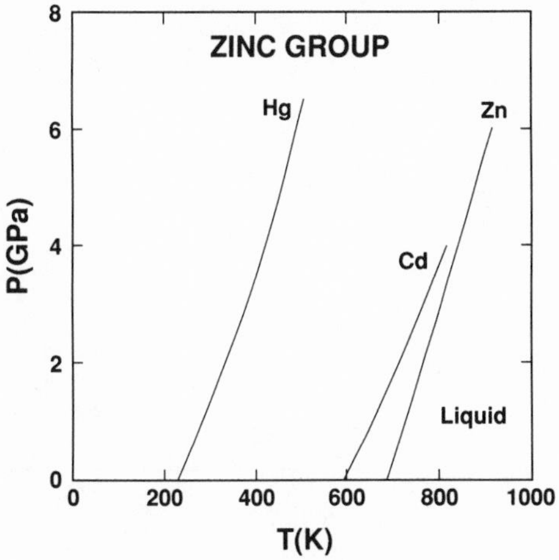

Fig. 13.14 The melting curves of the zinc group.

below ca. 50 K by straining the solid[12,125]. The structure of γ-Hg has not been determined.

The melting curve of Hg is an important pressure standard, and it has been determined to 6.5 GPa[123]. The phase diagram of Hg is shown in Fig. 13.15.

13.11.4 Theoretical

Although the zinc group metals are usually considered NFE metals, they are here included in the transition series for convenience. In fact, the filled *d* band does have an important effect on the structure of Zn, as determined by perturbation theories[126,127]. According to GPT calculations, the *d* band is responsible for the anomalous c/a ratio in Zn[126,128].

A detailed analysis of Hg phase stability at 0 K has been performed with the GPT[129]. This work clearly shows that fcc and hcp are mechanically unstable, and that the tetragonal and rhombohedral distortions of fcc corresponding to the β and α structures are stabilized. In addition, the hcp lattice is predicted to have a larger than ideal c/a ratio, as in Zn and Cd. A comparison of total energies upon compression shows β-Hg stable from

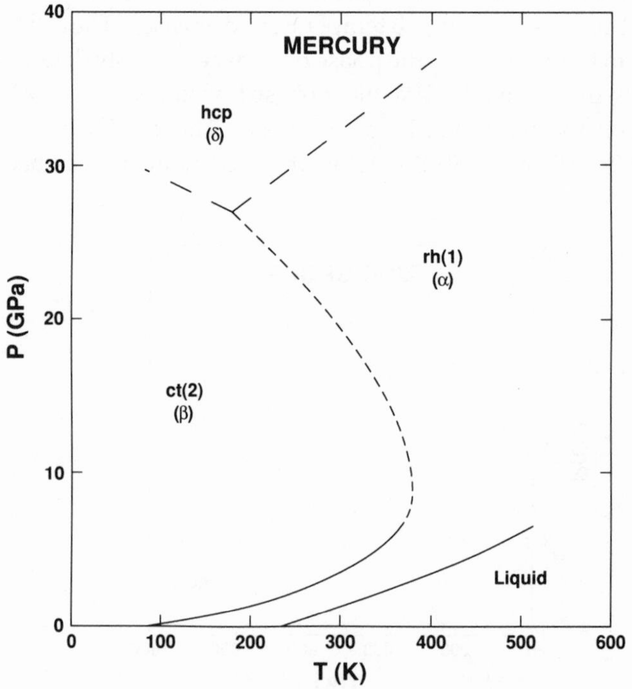

Fig. 13.15 The phase diagram of mercury.

0 to 100 GPa, where a transition to hcp is predicted. The hcp phase should remain stable up to 1 TPa. The overall agreement between experiment and theory is remarkably good.

13.12 Discussion

The transition metals are characterized by high densities, cohesive energies, and bulk moduli. These properties are the result of strong *d*-electron bonding. Plots of V_o, E_{coh}, and B_o show roughly symmetric curves with extreme values at the midpoint of the series, as illustrated in Fig. 13.16. The exception to this rule occurs in the 3*d* magnetic elements. This general behavior may be explained in terms of the simple Friedel model of the transition metal *d* bands[130]. Here the *d* band is represented by a rectangular density of states with energy E_B and width W_d and centered on the atomic energy E_d. The cohesive energy for a metal with n_d electrons is then

$$E_{coh} = E_d n_d - E_B = n_d W_d/2 - n_d^2 W_d/20 . \qquad (13.1)$$

This function has a parabolic form, as observed, and the increased bonding at the center of the *d* series increases the density and bulk modulus there also, as is observed. With the exception of the magnetic elements, modern band-structure calculations can accurately reproduce these properties of the transition metals[30,44,131,132].

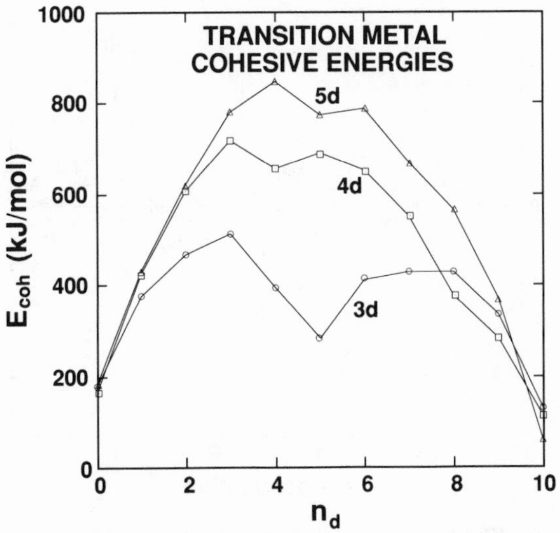

Fig 13.16 Transition-metal cohesive energies.

Theoretical discussion of the transition-metal crystal structures began in the 1930s with the work of Hume-Rothery[133]. This involved an application of the concepts of valence electrons and electronegativity to crystal stability. These ideas were extended by Engel and Brewer to form a correlation between *sp*-electron number and crystal structure[134].

This empirical approach has now been superseded by the more rigorous band-structure method[8,9]. Pettifor used a tight-binding orbital calculation to show that the structure sequence across the series was the result of the filling of the *d* band, and that the *sp*-electron number is nearly constant. This model correctly predicts the sequence hcp-bcc-hcp-fcc, although not all of the elements are correctly predicted[8].

Pettifor performed a canonical band-structure calculation, in which hybridization is ignored. If a fully self-consistent LMTO calculation is made, the agreement between theory and experiment is further improved and only Au and the magnetic elements Mn, Fe, and Co are incorrectly predicted[9]. The LMTO relative energies of the hcp, bcc, and fcc structures are shown in Fig. 13.17 for the 4*d* elements. The GPT method has now been extended to partially filled *d*-band metals, and it also predicts the observed structural trend[44]. Although the theories correctly predict the structural trend in the transition metals, alloy data suggest that the theoretical energy

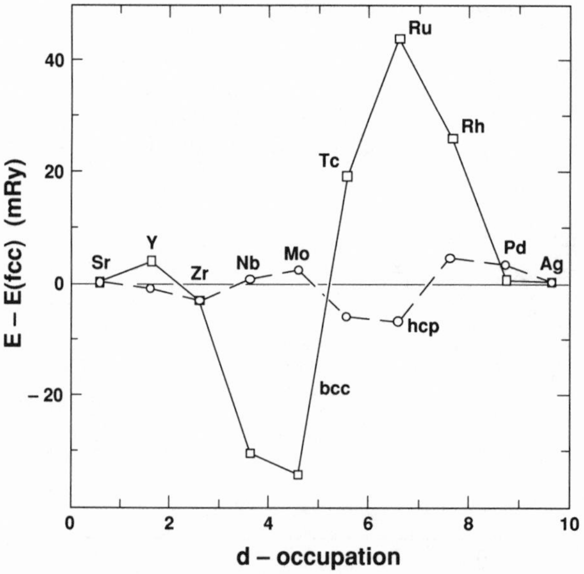

Fig. 13.17 LMTO structural energies for the 4*d* transition metals. (From Skriver[9]. Redrawn with permission.)

differences between structures are too large, possibly indicating a weakness in the local-density approximation[9].

Magnetism in the *3d* elements remains a challenge for theory. Ferromagnetism occurs when there are narrow *d* bands and sharp peaks in the density of states near the Fermi level. The increased kinetic energy associated with these peaks is compensated by a lower exchange energy for the magnetically ordered state. The accurate calculation of the magnetic ground state remains difficult, as in the case of Fe. The local spin-density approximation is the prevailing theoretical model, but it is not sufficient for accurate total energies. Nonlocal-density theories appear to be required.

A summary of the static high pressures achieved in transition metals is shown in Fig. 13.18. Because the transition metals are relatively incompressible, little volume compression occurs under normal DAC pressures, and phase transitions are therefore not observed. At very high pressures ($P >$ 100 GPa), we fully expect to see phase changes induced by changes in the band structure, but current experimental limitations make their observation difficult. At the same time, there has not yet been a systematic theoretical

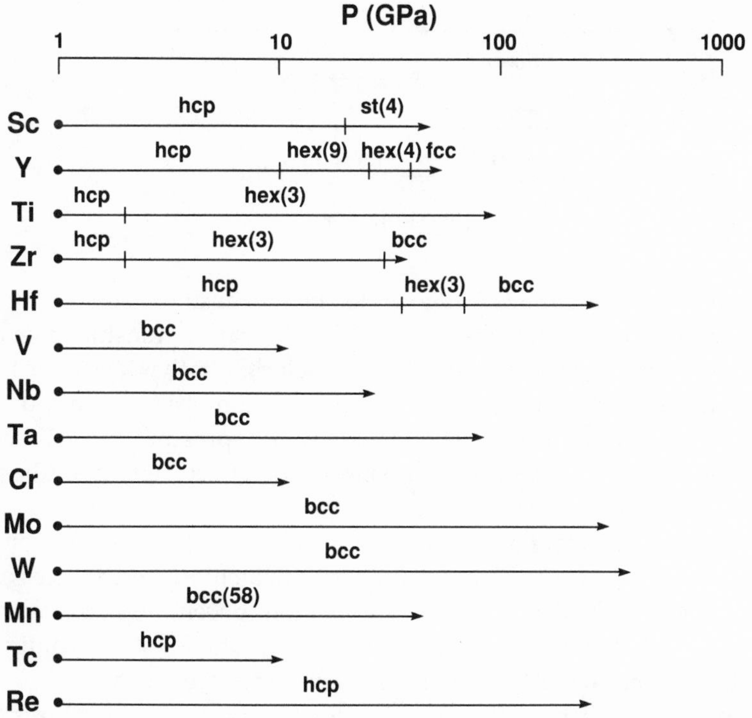

Fig. 13.18 Summary of RT high-pressure structures in the transition metals.
(continued)

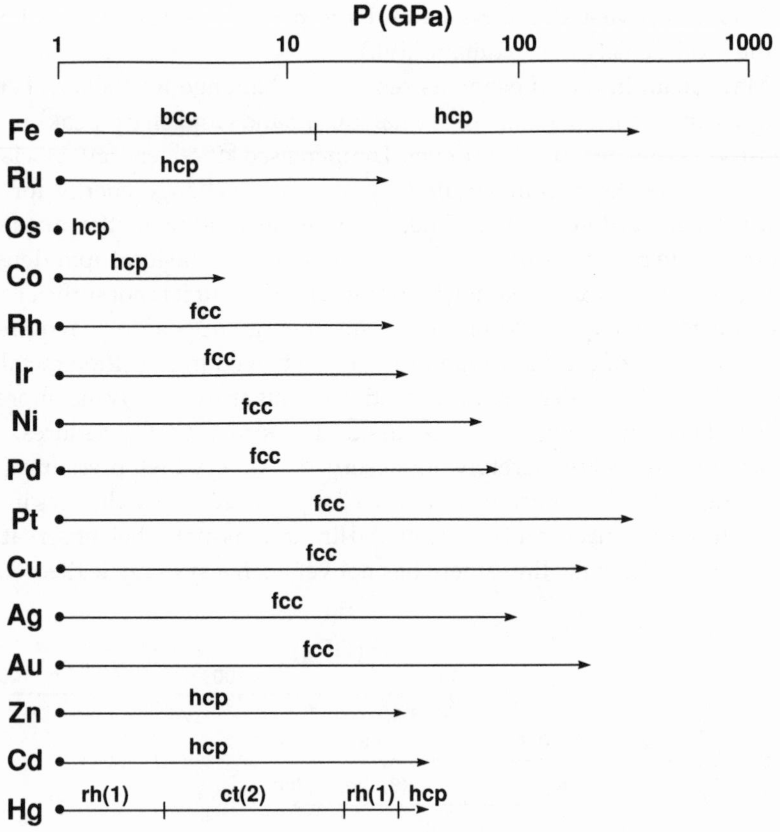

Fig. 13.18 (*continued*).

study of the phase transitions expected in the transition metals at ultrahigh pressures. In general, it is expected that the earlier transition metals will take on the structures of the elements to their right as the *s*-*d* electron transfer fills the *d* band under pressure. The transition observed in Mo at high pressure may be an example of this if the high-pressure phase is hcp. For the later members of the series, pressure has the effect of emptying the *d* band, so the reverse trend is expected.

Transition-metal phase transitions can also be driven by alloying. If a transition metal such as Mo is alloyed with elements richer in *d* electrons, the bcc phase will be destabilized at a *d*-electron concentration corresponding well with the band-structure predictions[135]. This is a strong confirmation of the *d*-electron picture of the transition-metal structures.

For the prediction of phase changes at nonzero temperatures, static-lattice band-structure calculations are no longer useful. The GPT seems to be the appropriate theory for this purpose. For partially filled *d*-bands, both two-body and three-body forces are important in determining structure, but higher terms in the energy expansion are very small. Tests of GPT phonon-dispersion curves against experimental data so far are encouraging[44,136]. The GPT may thus be used in lattice dynamics and in liquid theory to compute solid-solid and solid-liquid phase boundaries, and it has achieved some initial successes in this direction.

References

1. W. A. Grosshans, Y. K. Vohra, and W. B. Holzapfel, J. Magn. Magn. Mat. **29**, 282 (1982).
2. Y. K. Vohra, W. Grosshans, and W. B. Holzapfel, Phys. Rev. B **25**, 6019 (1982).
3. J. Akella, J. Xu, and G. S. Smith, Physica **139&140B** 285 (1986).
4. W. J. Carter, J. N. Fritz, S. P. Marsh, and R. G. McQueen, J. Phys. Chem. Solids **36**, 741 (1975).
5. R. Hultgren, P. D. Desai, D. T. Hawkins, M. Gleiser, K. K. Kelley, and D. D. Wagman, *Selected Values of the Thermodynamic Properties of the Elements* (American Society for Metals, Metals Park, Ohio, 1973).
6. Y. K. Vohra, H. Olijnyk, W. Grosshans, and W. B. Holzapfel, Phys. Rev. Lett. **47**, 1065 (1981).
7. W. A. Grosshans, Y. K. Vohra, and W. B. Holzapfel, Phys. Rev. Lett. **49**, 1572 (1982).
8. D. G. Pettifor, CALPHAD **1**, 305 (1977).
9. H. L. Skriver, Phys. Rev. B **31**, 1909 (1985).
10. B. J. Min and K.-M. Ho, Phys. Rev. B **40**, 7532 (1989).
11. A. K. McMahan, H. L. Skriver, and B. Johansson, Phys. Rev. B **23**, 5016 (1981).
12. J. Donohue, *The Structures of the Elements* (Wiley, New York, 1974) chap. 6.
13. S. K. Sikka, Y. K. Vohra, and R. Chidambaram, Prog. Mat. Sci. **27**, 245 (1982).
14. F. P. Bundy, General Electric Report No. 63-RL-3481C, Oct. 1963.
15. V. A. Zil'bershteyn, G. I. Nosova, and E. I. Estrin, Fiz. Metal. Metalloved. **35**, 584 (1973) [Phys. Metal. Metallog. **35**, 128 (1973)].
16. V. A. Zil'bershteyn, N. P. Chistotina, A. A. Zharov, N. S. Grishina, and E. I. Estrin, Fiz. Metal. Metalloved. **39**, 445 (1975) [Phys. Metal. Metallog. **39**, 208 (1975)].
17. Y. K. Vohra, H. Olijnyk, W. Grosshans, and W. B. Holzapfel, in *High Pressure in Research and Industry*, C.-M. Backmann, T. Johannisson, and L. Tegnér, eds. (Arkitektkopia, Uppsala, 1982) p. 354.
18. H. Xia, G. Parthasarathy, H. Luo, Y. K. Vohra, and A. L. Ruoff, Phys. Rev. B **42**, 6736 (1990).
19. A. Jayaraman, W. Klement, Jr., and G. C. Kennedy, Phys. Rev. **131**, 644 (1963).
20. R. G. McQueen, S. P. Marsh, J. W. Taylor, J. N. Fritz, and W. J. Carter, in *High-Velocity Impact Phenomena*, R. Kinslow, ed. (Academic, New York, 1970) p. 344.
21. A. R. Kutsar, M. N. Pavlovskii, and V. V. Komissarov, Pis'ma Zh. Eksp. Teor. Fiz. **35**, 91 (1982) [JETP Lett. **35**, 108 (1982)].

22. A. N. Kiselev and A. A. Fal'kov, Fiz. Goren. Vzryva **18**, 115 (1982) [Combustion, Explosion, Shock Wave **18**, 94 (1982)].
23. C. E. Morris, personal communication.
24. G. R. Gathers, Int. J. Thermophys. **4**, 271 (1983).
25. A. Fernandez Guillermet, High Temp.-High Press. **19**, 119 (1987).
26. H. Xia, S. J. Duclos, A. L. Ruoff, and Y. K. Vohra, Phys. Rev. Lett. **64**, 204 (1990).
27. A. R. Kutsar, M. N. Pavlovskii, and V. V. Komissarov, Pis´ma Zh. Eksp. Teor. Fiz. **39**, 399 (1984) [JETP Lett. **39**, 480 (1984)].
28. J. S. Gyanchandani, S. C. Gupta, S. K. Sikka, and R. Chidambaram, in *Shock Compression of Condensed Matter—1989*, S. C. Schmidt, J. N. Johnson, and L. W. Davison, eds. (North-Holland, Amsterdam, 1990) p. 131.
29. J. S. Gyanchandani, S. C. Gupta, S. K. Sikka, and R. Chidambaram, J. Phys.: Condens. Matter **2**, 6457 (1990).
30. G. W. Fernando, R. E. Watson, M. Weinert, Y. J. Wang, and J. W. Davenport, Phys. Rev. B **41**, 11813 (1990).
31. H. W. Myron, A. J. Freeman, and S. C. Moss, Solid State Comm. **17**, 1467 (1975).
32. A. L. Simons and C. M. Varma, Solid State Comm. **35**, 317 (1980).
33. L. Ming and M. H. Manghnani, J. Appl. Phys. **49**, 208 (1978).
34. G. R. Gathers, J. Appl. Phys. **59**, 3291 (1986).
35. G. R. Gathers, J. W. Shaner, R. S. Hixson, and D. A. Young, High Temp.-High Press. **11**, 653 (1979).
36. L. F. Vereshchagin, A. A. Semerchan, N. N. Kuzin, and S. V. Popova, Dokl. Akad. Nauk SSSR **138**, 84 (1961) [Sov. Phys. Dokl. **6**, 391 (1961)].
37. S. P. Marsh, ed., *LASL Shock Hugoniot Data* (University of California Press, Berkeley, Los Angeles, London, 1980).
38. J. W. Shaner, G. R. Gathers, and W. M. Hodgson, in *Proceedings of the Seventh Symposium on Thermophysical Properties*, A. Cezairliyan, ed. (AMSE, New York, 1977) p. 896.
39. J. Xu, H.-K. Mao, and P. M. Bell, High Temp.-High Press. **16**, 495 (1984).
40. A. C. Mitchell and W. J. Nellis, J. Appl. Phys. **52**, 3363 (1981).
41. N. S. Fateeva and L. F. Vereshchagin, Dokl. Akad. Nauk SSSR **197**, 1060 (1971) [Sov. Phys. Dokl. **16**, 322 (1971)].
42. J. W. Shaner, G. R. Gathers, and C. Minichino, High Temp.-High Press. **9**, 331 (1977).
43. J. M. Brown and J. W. Shaner, in *Shock Waves in Condensed Matter—1983*, J. R. Asay, R. A. Graham, and G. K. Straub, eds. (North-Holland, Amsterdam, 1984) p. 91.
44. J. A. Moriarty, Phys. Rev. Lett. **55**, 1502 (1985).
45. M. O. Steinitz, L. H. Schwartz, J. A. Marcus, E. Fawcett, and W. A. Reed, Phys. Rev. Lett. **23**, 979 (1969).
46. H. Umebayashi, G. Shirane, B. C. Frazer, and W. B. Daniels, J. Phys. Soc. Japan **24**, 368 (1968).
47. T. Mitsui and C. T. Tomizuka, Phys. Rev. **137**, A564 (1965).
48. W. E. Evenson and H. T. Hall, Science **150**, 1164 (1965).
49. H. K. Mao, P. M. Bell, J. W. Shaner, and D. J. Steinberg, J. Appl. Phys. **49**, 3276 (1978).
50. Y. K. Vohra and L. Ruoff, Bull. Am. Phys. Soc. **35**, 808 (1990).

51. R. S. Hixson, D. Boness, J. W. Shaner, and J. A. Moriarty, Phys. Rev. Lett. **62**, 637 (1989).
52. N. F. Fateeva and L. F. Vereshchagin, ZhETF Pis. Red. **14**, 233 (1971) [JETP Lett. **14**, 153 (1971)].
53. A. Fernandez Guillermet, Int. J. Thermophys. **6**, 367 (1985).
54. A. L. Ruoff, H. Xia, H. Luo, and Y. K. Vohra, Appl. Phys. Lett. **57**, 1007 (1990).
55. R. Hixson, personal communication.
56. L. F. Vereshchagin and N. S. Fateeva, Zh. Eksp. Teor. Fiz. **55**, 1145 (1968) [JETP **28**, 597 (1969)].
57. R. S. Hixson and M. A. Winkler, preprint.
58. L. F. Mattheiss and D. R. Hamann, Phys. Rev. B **33**, 823 (1986).
59. C. T. Chan, D. Vanderbilt, S. G. Louie, and J. R. Chelikowsky, Phys. Rev. B **33**, 7941 (1986).
60. J. Chen, D. Singh, and H. Krakauer, Phys. Rev. B **38**, 12834 (1988).
61. J. A. Moriarty, personal communication.
62. K. Takemura, O. Shimomura, K. Hase, and T. Kikegawa, J. Phys. F **18**, 197 (1988).
63. E. Rapoport and G. C. Kennedy, J. Phys. Chem. Solids **27**, 93 (1966).
64. P. W. Bridgman, Proc. Am. Acad. Arts Sci. **84**, 117 (1955).
65. Y. K. Vohra, S. J. Duclos, and A. L. Ruoff, Phys. Rev. B **36**, 9790 (1987).
66. Y. K. Vohra and A. L. Ruoff, High Press. Res. **4**, 296 (1990).
67. L. F. Vereshchagin, N. S. Fateeva, and M. V. Magnitskaya, ZhETF Pis. Red. **22**, 229 (1975) [JETP Lett. **22**, 106 (1975)].
68. R. E. Watson, J. W. Davenport, M. Weinert, and G. Fernando, Phys. Rev. B **38**, 7817 (1988).
69. S. Akimoto, T. Suzuki, T. Yagi, and O. Shimomura, in *High Pressure Research in Mineral Physics*, M. H. Manghnani and Y. Syono, eds. (Terra Scientific, Tokyo, 1987) p. 149.
70. M. H. Manghnani, L. C. Ming, and N. Nakagiri, in Ref. 69, p. 155.
71. E. Huang, W. A. Bassett, and P. Tao, in Ref. 69, p. 165.
72. R. Boehler, N. von Bargen, and A. Chopelas, J. Geophys. Res. **95**, 21731 (1990).
73. G. Cort, R. D. Taylor, and J. O. Willis, J. Appl. Phys. **53**, 2064 (1982).
74. A. P. Jephcoat, H. K. Mao, and P. M. Bell, J. Geophys. Res. **91**, 4677 (1986).
75. H. K. Mao, Y. Wu, L. C. Chen, J. F. Shu, and R. J. Hemley, High Press. Res. **5**, 773 (1990).
76. A. Fernandez Guillermet and P. Gustafson, High Temp.-High Press. **16**, 591 (1985).
77. F. P. Bundy, J. Appl. Phys. **36**, 616 (1965).
78. A. V. Omel´chenko, V. I. Soshnikov, and E. I. Estrin, Fiz. Metal. Metalloved. **28**, 77 (1969) [Phys. Met. Metallog. **28**, 80 (1969)].
79. P. W. Mirwald and G. C. Kennedy, J. Geophys. Res. **84**, 656 (1979).
80. H. K. Mao, P. M. Bell, and C. Hadidiacos, in Ref. 69, p. 135.
81. R. Boehler, Geophys. Res. Lett. **13**, 1153 (1986).
82. J. M. Brown and R. G. McQueen, J. Geophys. Res. **91**, 7485 (1986).
83. M. Ross, D. A. Young, and R. Grover, J. Geophys. Res. **95**, 21713 (1990).
84. H. M. Strong, R. E. Tuft, and R. E. Hanneman, Metal. Trans. **4**, 2657 (1973).
85. L.-G. Liu and W. A. Bassett, J. Geophys. Res. **80**, 3777 (1975).

86. Q. Williams, R. Jeanloz, J. Bass, B. Svendsen, and T. J. Ahrens, Science **236**, 181 (1987).
87. J. D. Bass, in Ref. 69, p. 393.
88. R. L. Clendenen and H. G. Drickamer, J. Phys. Chem. Solids **25**, 865 (1964).
89. C. S. Wang, B. M. Klein, and H. Krakauer, Phys. Rev. Lett. **54**, 1852 (1985).
90. H. J. F. Jansen and S. S. Peng, Phys. Rev. B **37**, 2689 (1988).
91. H. S. Greenside and M. A. Schlüter, Phys. Rev B **27**, 3111 (1983).
92. P. Bagno, O. Jepsen, and O. Gunnarsson, Phys. Rev. B **40**, 1997 (1989).
93. H. Hasegawa and D. G. Pettifor, Phys. Rev. Lett. **50**, 130 (1983).
94. G. Grimvall, Solid State Comm. **14**, 551 (1974).
95. O. L. Anderson, Geophys. J. Roy. Astr. Soc. **84**, 561 (1986).
96. F. Mulargia and E. Boschi, in *Physics of Earth's Interior*, A. M. Dziewonski and E. Boschi, eds. (North-Holland, Amsterdam, 1980) p. 337.
97. D. A. Young and R. Grover, in Ref. 43, p. 65.
98. C. Hausleitner and J. Hafner, J. Phys.: Condens. Matter **1**, 5243 (1989).
99. A. Fernandez Guillermet, Int. J. Thermophys. **8**, 481 (1987).
100. G. C. Kennedy and R. C. Newton, in *Solids under Pressure*, W. Paul and D. W. Warschauer, eds. (McGraw-Hill, New York, 1963) chap. 7.
101. T. Nishizawa and K. Ishida, Bull. Alloy Phase Diagr. **4**, 387 (1983).
102. E. A. Perez-Albuerne, K. F. Forsgren, and H. G. Drickamer, Rev. Sci. Instrum. **35**, 29 (1964).
103. H. M. Strong and F. P. Bundy, Phys. Rev. **115**, 278 (1959).
104. J. Akella, J. Phys. Chem. Solids **43**, 941 (1982).
105. B. I. Min, T. Oguchi, and A. J. Freeman, Phys. Rev. B **33**, 7852 (1986).
106. J. Akella, personal communication.
107. N. C. Holmes, J. A. Moriarty, G. R. Gathers, and W. J. Nellis, J. Appl. Phys. **66**, 2962 (1989).
108. G. M. Gandel'man, V. M. Ermachenko, and Ya. B. Zel'dovich, Zh. Eksp. Teor. Fiz. **44**, 386 (1963) [JETP **17**, 263 (1963)].
109. A. K. McMahan and R. C. Albers, Phys. Rev. Lett. **49**, 1198 (1982).
110. A. K. McMahan, personal communication.
111. P. M. Bell, J. Xu, and H. K. Mao, in *Shock Waves in Condensed Matter*, Y. M. Gupta, ed. (Plenum, New York, 1986) p. 125.
112. P. W. Mirwald and G. C. Kennedy, J. Geophys. Res. **84**, 6750 (1979).
113. L.-G. Liu and W. A. Bassett, J. Appl. Phys. **44**, 1475 (1973).
114. D. L. Heinz and R. Jeanloz, J. Appl. Phys. **55**, 885 (1984).
115. L. C. Ming, D. Xiong, and M. H. Manghnani, Physica **139&140B**, 174 (1986).
116. B. K. Godwal and R. Jeanloz, Phys. Rev. B **40**, 7501 (1989).
117. N. Takeuchi, C. T. Chan, and K. M. Ho, Phys. Rev. B **40**, 1565 (1989).
118. W. J. Nellis, J. A. Moriarty, A. C. Mitchell, M. Ross, R. G. Dandrea, N. W. Ashcroft, N. C. Holmes, and G. R. Gathers, Phys. Rev. Lett. **60**, 1414 (1988).
119. J. A. Moriarty, in Ref. 111, p. 101.
120. O. Schulte, A. Nikolaenko, and W. B. Holzapfel, preprint.
121. J. Akella, J. Ganguly, R. Grover, and G. Kennedy, J. Phys. Chem. Solids **34**, 631 (1973).
122. C. W. F. T. Pistorius, High Temp.-High Press. **7**, 451 (1975).
123. W. Klement, Jr., A. Jayaraman, and G. C. Kennedy, Phys. Rev. **131**, 1 (1963).

124. O. Schulte and W. B. Holzapfel, Phys. Lett. A **131**, 38 (1988).
125. J. S. Abell, A. G. Crocker, and H. W. King, Phil. Mag. **21**, 207 (1970).
126. J. A. Moriarty, Int. J. Quantum Chem. Symp. **17**, 541 (1983).
127. M. Hasegawa, M. Moriyama, M. Watabe, and W. H. Young, J. Phys. Soc. Jpn. **57**, 1308 (1988).
128. J. A. Moriarty, Phys. Rev. B **26**, 1754 (1982).
129. J. A. Moriarty, Phys. Lett. A **131**, 41 (1988).
130. W. A. Harrison, *Electronic Structure and the Properties of Solids* (W. H. Freeman, San Francisco, 1980) chap. 20.
131. V. L. Moruzzi, J. F. Janak, and A. R. Williams, *Calculated Electronic Properties of Metals* (Pergamon, New York, 1978).
132. D. Glötzel, in *Physics of Solids under High Pressure*, J. S. Schilling and R. N. Shelton, eds. (North-Holland, Amsterdam, 1981) p. 263.
133. W. Hume-Rothery and G. V. Raynor, *The Structure of Metals and Alloys* 4th ed. (Institute of Metals, London, 1962).
134. L. Brewer, Science **161**, 115 (1968).
135. J. W. Shaner, in Ref. 28, p. 121.
136. J. A. Moriarty, Phys. Rev. B **38**, 3199 (1988).

CHAPTER 14
The Lanthanides

14.1 Introduction

The lanthanides include lanthanum plus the 14 following elements in which the 4*f* electron shell is filled. Except for the two divalent metals Eu and Yb, all of these elements are trivalent, similar to the transition metals Sc and Y.

Because the 4*f* electrons are deeply buried in the lanthanide atoms, they are localized and do not participate in metallic bonding[1]. The *spd* valence electrons dominate the bonding, and this permits the phase diagrams to vary slowly across the series so that their relationship to one another is clearly seen[2]. The continuity of the lanthanide properties is more obvious, for example, than those of the transition metals, where the *d* electrons are important in the bonding. Also, the actinides form an analogous series in which the 5*f* shell is filled, and comparison of the phase diagrams of the two series is useful and informative.

The most characteristic fact about the lanthanides is the sequence of close-packed phases which can be seen either by changing the atomic number Z at RP or by isothermal compression of a single element[2]. With increasing pressure, the sequence is hcp (hex(2))→Sm-type (rh(3) or hex(9))→dhcp (hex(4))→fcc (fcc(4)). These phases may be represented[3] by the uniquely positioned hexagonal layer planes A, B, and C. Thus hcp = ABABAB..., Sm-type = ABACACBCB..., dhcp = ABAC..., and fcc = ABCABC.... A more compact notation is to represent a layer plane as either hexagonal (h) if its two neighboring layers are equivalent, or cubic (c) if they are inequivalent. Thus hcp = hhh..., Sm-type = hhchhc..., dhcp = hchchc..., and fcc = ccc.... A new high-pressure phase referred to as "distorted fcc" has been found beyond fcc[4]. This phase has a six-layer structure, and there is currently a dispute about the correct space group for this phase, which may have spatial modulations in the interlayer distances (distorted fcc), may be "triple hexagonal close-packed" (thcp), or may be another structure[5–8]. In this chapter the lanthanide phases will be referred to as hcp, hex(9)

(Sm-type), hex(4) (dhcp), fcc, and hex(6) ("distorted fcc") for convenience. The four common lanthanide structures are shown in Fig. 14.1. Since there is an infinite number of close-packed structures between fcc and hcp, a complete theory of the lanthanide structures must show why only the observed phases are stable. These phase transitions are sluggish and the equilibrium pressures are not precisely determined.

14.2 Lanthanum

At RTP La is hex(4). Compression at RT to ca. 1.9 GPa leads to a transition to fcc[8–10]. Further compression to 7.0 GPa leads to hex(6)[4,5,11]. No further transitions have been observed up to 40 GPa[11].

The hex(4)-fcc phase boundary has a negative dP/dT and appears to intersect the $T = 0$ axis at ca. 2.0 GPa[8,10], and the $P = 0$ axis at 550 K[12]. The fcc-hex(6) phase boundary has been measured to 450 K and 14 GPa, and has a positive dP/dT[11].

At RP and 1134 K, there is a transition to bcc[12]. The bcc phase disappears in a hex(4)-bcc-liquid triple point at 2.0 GPa[13]. The melting curve has been measured to 3.5 GPa[13]. The La phase diagram is shown in Fig. 14.2.

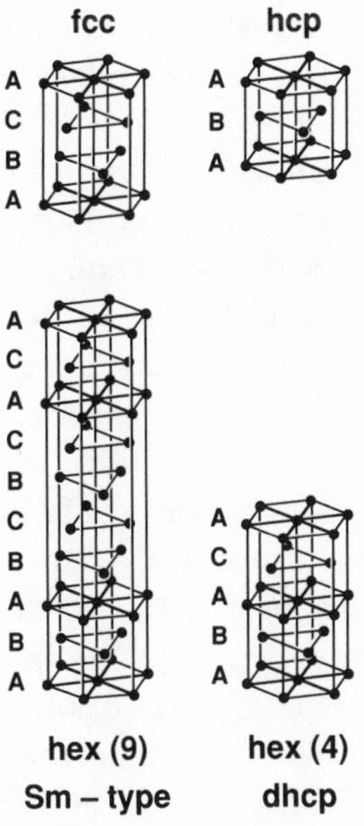

Fig. 14.1 The crystal structures of the lanthanides. (From Alstetter[3]. Redrawn with permission.)

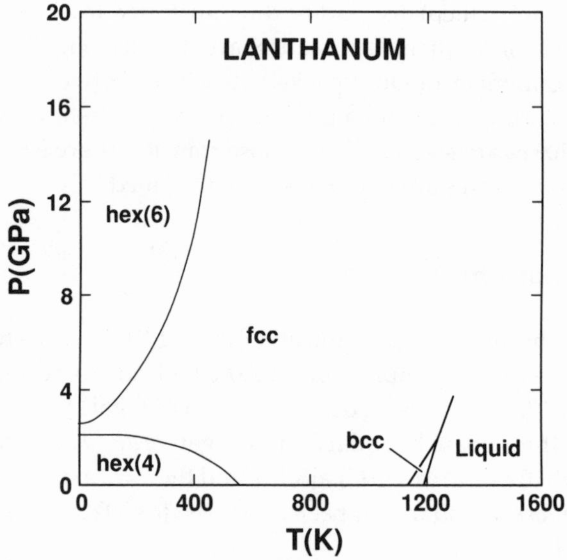

Fig. 14.2 The phase diagram of lanthanum.

La has been studied with unusual thoroughness at high pressure. There have been measurements of superconducting T_c[14] and sound speed[15] under static conditions, and sound speed under shock loading[16]. All of these measurements show interesting anomalies corresponding to the observed phase transitions and in general to the *s-d* electron-transfer process.

LMTO calculations have provided a rather complete picture of the thermodynamic behavior of La at high pressure[17,18]. The theory correctly predicts the hex(4)-fcc transition, although at ca. 9 GPa, which is too high[18]. In the range $0.50 < V/V_0 < 0.57$ the *P-V* isotherm steepens dramatically[17]. This corresponds to the rapid variation in 6*s*-electron occupation. Under compression the 6*s* band rises above the 5*d* band. The *s-d* electron transfer occurs first by hybridization, and then by a rapid emptying of the 6*s* band. High-temperature LMTO calculations show that the *s-d* effect is weakened by temperature, but that there is a correlated anomaly in the Grüneisen parameter γ due to the rapid change in phonon frequencies in the *s-d* transfer region[17]. The sharp peak predicted in the γ function is instrumental in producing the observed kink in the u_s-u_p curve of the shock Hugoniot[19]. This excellent agreement between theory and experiment for the La shock Hugoniot is shown in Fig. 14.3. Previous theories about this feature of the lanthanides being caused by melting[19] or xenon-core overlap[20] are thus in error.

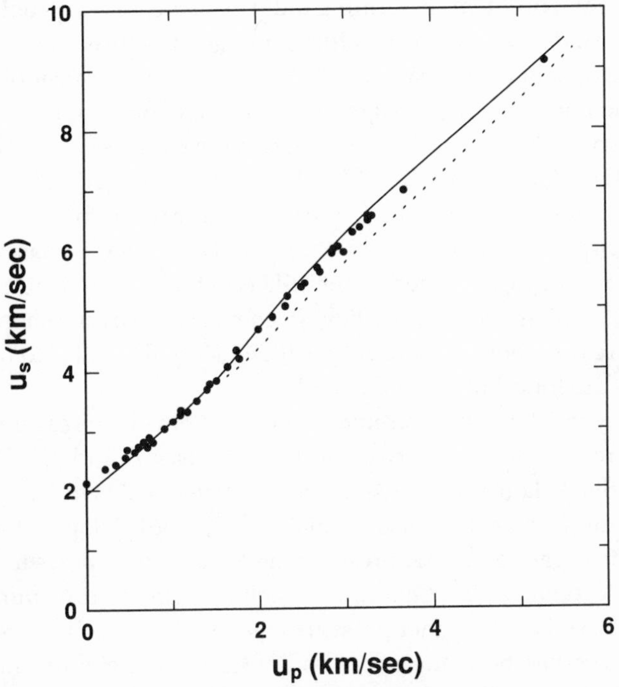

Fig. 14.3 Experimental and theoretical shock-Hugoniot data for lanthanum. The points are experimental shock data. The dotted line is the theory without the *s-d* transfer, and the solid curve is the theory with the *s-d* transfer.

LAPW total-energy calculations on fcc and bcc La at 0 K show that the bcc energy is higher, as observed, and that the energy difference is in rough accord with the entropy change observed at the fcc-bcc transition point at RP[21].

14.3 Cerium

Of all the lanthanides, Ce has attracted the most attention both experimentally and theoretically because of its unusual phase diagram. At RP, Ce has 4 solid phases: for $T < 96$ K, α fcc; for $96 < T < 326$ K, β hex(4); for $326 < T < 999$ K, γ fcc; and for $999 < T < 1071 \text{ K} = T_m$, δ bcc[12,22]. Under compression, the hex(4) phase disappears and there is a first-order phase transition from γ to α. This isostructural transition occurs with a 17% volume change at RT and 0.7 GPa[22,23]. Even more remarkable than the isostructural transition itself is the termination of this transition in a critical point near 550 K and 1.75 GPa[24]. The precise location of the critical point is disputed because of experimental problems in locating it[25]. The theoretical problem posed by this unusual behavior has produced a large literature.

Compression of the α phase at RT to 5.1 GPa gives a transition to a new phase, which was initially designated α′[26]. A contentious dispute has developed over the structure of this phase, with claims for fcc[26], hcp[27], eco(4) (α-U)[28], and cm(2)[28,29]. The most recent study[29] favors the cm(2) phase, now designated α″. Further compression to 12.2 GPa yields still another phase (ε), found to be ct(2)[30]. The α″ and ε phases are both slight distortions of fcc, whereas the α-U eco(4) structure is not. Both the α-α″ and α″-ε phase boundaries have been measured over a short range of temperatures and both have negative dP/dT slopes[31,32]. No further RT transitions are found up to 46 GPa[29].

Extensive resistance measurements on Ce at both high pressure and high temperature have suggested two additional phases, β′ and γ′[33]. Confirmation of these claims will have to await detailed XRD work.

The melting curve of Ce is also anomalous[24]. The RP slope of the δ-liquid boundary is negative. Under pressure the δ phase disappears in a fcc-bcc-liquid triple point at 2.6 GPa and a melting temperature minimum is reached at 3.3 GPa. At higher pressures the melting slope is positive. The melting curve has been measured to 7.0 GPa. It is significant that the extrapolated γ-α, α-α″, and α″-ε phase boundaries all meet at the melting minimum. The phase diagram of Ce is shown in Fig. 14.4.

Theoretical work on Ce has concentrated on the γ-α isostructural transition and on the melting curve. The first theoretical explanation for the transition was a promotion of the atomic 4*f* electron to the (5*d*6*s*) conduction band[22,34]. Such a promotion would mean a valence change from $Z_i = 3$ to $Z_i = 4$, and a consequent volume collapse, as observed. This model requires a 4*f* level and (5*d*6*s*) valence band which are nearly degenerate. Subsequent experimental work, mainly spectroscopic, has effectively refuted this idea. The valence does not change appreciably across the transition[35].

LMTO and FPLAPW calculations in which the 4*f* electrons are treated as atomic-like or bandlike predict a large volume difference between these approximations, suggesting that the γ-α transition may be of the localization-delocalization type[36,37]. Standard one-electron models like LMTO and FPLAPW do not in their unmodified form predict any anomaly in Ce under pressure, which shows that the γ-α transition is driven by correlation-exchange effects not treated by these models[37]. A large number of suggestions have been made about the physics involved: Mott transition[38], *d-f* hybridization gap[39,40], Anderson mixing[41], Jahn-Teller effect[42], Kondo effect[43,44], Fermi liquid[45], and

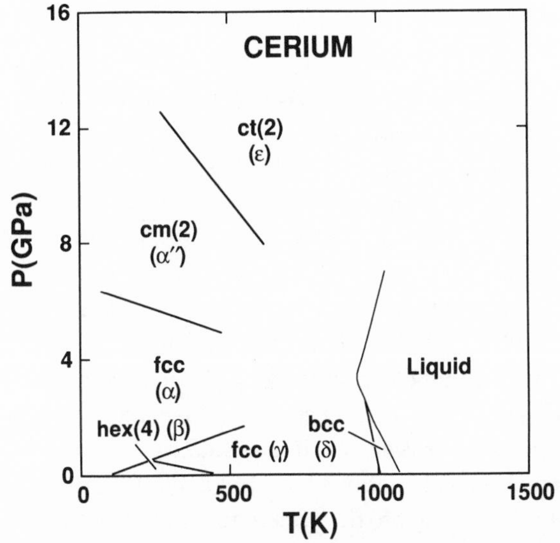

Fig. 14.4 The phase diagram of cerium.

interconfigurational mixing or mixed valence[46]. All of these models require fitted parameters in order to reproduce the observed γ-α transition. Several of the models have been compared in detail with the γ-α transition, and they include the critical point. The Kondo-effect model[43] predicts two critical points, one at $P < 0$. This feature is observed in certain Ce alloys[45]. There is as yet no agreement on the physics of the γ-α transition.

LMTO structural calculations with bandlike 4*f* electrons predict a transition from fcc to a ct(2) tetragonal distortion under pressure, in agreement with experiment[18].

The melting curve of Ce has been explained phenomenologically in terms of a pseudobinary alloy consisting of a mixture of large (γ) and small (α) atoms[47]. At low pressures, the low-density (γ) solid is in equilibrium with a liquid mixture of γ and α atoms at higher density, yielding $\Delta V_m < 0$ and $dP/dT < 0$. At higher pressure, the solid becomes pure α, with a higher density than the liquid mixture, giving $\Delta V_m > 0$ and $dP/dT > 0$. At the point between phase regions, $\Delta V_m = 0$ and $dP/dT = 0$, which is the melting-curve minimum. The pseudobinary model has also been applied to the solid γ-α transition, with good results[47]. Thus the thermodynamic behavior underlying the main features of the Ce phase diagram is well understood.

14.4 Praseodymium

At RTP, Pr is hex(4). Compression at RT to 3.8 GPa leads to an fcc phase[8,48]. At 6.2 GPa a hex(6) phase appears[49,50]. At ca. 20 GPa, the hex(6) structure transforms to the eco(4) α-U structure with a 9–10% volume change[49,50]. Compression to 40 GPa shows no further transition[49].

The hex(4)-fcc phase boundary has a negative slope, but there is conflicting evidence whether it intersects the $P = 0$ axis[51,52]. Here I assume that the phase boundary does intersect the $P = 0$ axis. The fcc-hex(6) phase boundary has a positive slope and there may be an hex(4)-fcc-hex(6) triple point below RT[8]. The hex(6)-eco(4) phase boundary also has a positive slope[53].

A bcc phase appears before melting. This phase disappears in a triple point at ca. 5 GPa[13]. The melting curve has been measured to 7.0 GPa[13]. The phase diagram of Pr is shown in Fig. 14.5.

Detailed structural calculations have not been performed on Pr, but it is clear that Pr fits smoothly into the systematics of the lanthanide series. The high-density ("collapsed") structure observed at 20 GPa is evidently an indication of pressure-induced 4*f* bonding, similar to that in α-Ce[49,54].

14.5 Neodymium

At RTP, Nd is hex(4). RT compression to ca. 5.8 GPa yields fcc[8,48,55]. Further compression causes transitions to hex(6) at ca. 18 GPa and to a

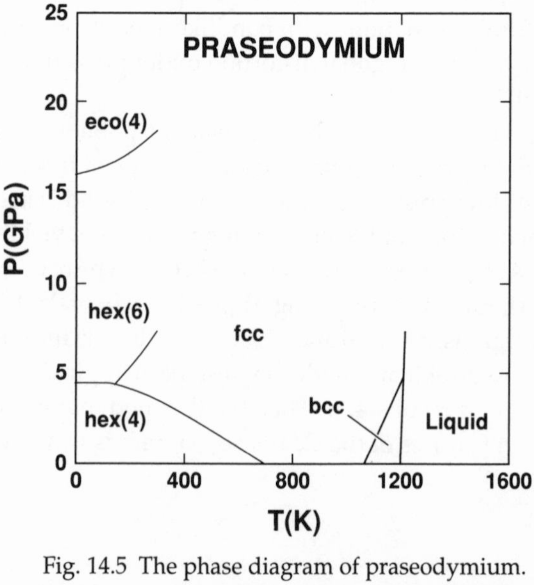

Fig. 14.5 The phase diagram of praseodymium.

lower-symmetry cm(4) phase at 41 GPa[11,56,57]. The cm(4) structure is considered to be an actinide-like phase analogous to the eco(4) phase in Pr[11,57]. Still further compression to 80.5 GPa yields evidence of a new, unidentified phase at ca. 70 GPa[58]. The hex(4)-fcc phase boundary has a negative slope[55] and evidently intersects the $P = 0$ axis[51,59].

At higher temperature, bcc becomes stable. The bcc-fcc boundary has been followed to 6.0 GPa[13]. The bcc phase will probably disappear in a triple point at some higher pressure. The melting curve has been measured to 6.0 GPa[13]. The phase diagram of Nd is shown in Fig. 14.6.

14.6 Promethium

Pm is an unstable element whose most readily available isotope has a half-life of 2.6 yr. The pure element has been prepared, and the RTP structure is hex(4)[60], as in Pr and Nd. DAC compression of Pm up to 60 GPa reveals the usual light-lanthanide sequence of transitions: hex(4)→fcc at ca. 10 GPa; fcc→hex(6) at ca. 18 GPa; and a possible further phase change above 40 GPa[61]. Pm melts at 1315 K[62]. The melting curve has not been measured.

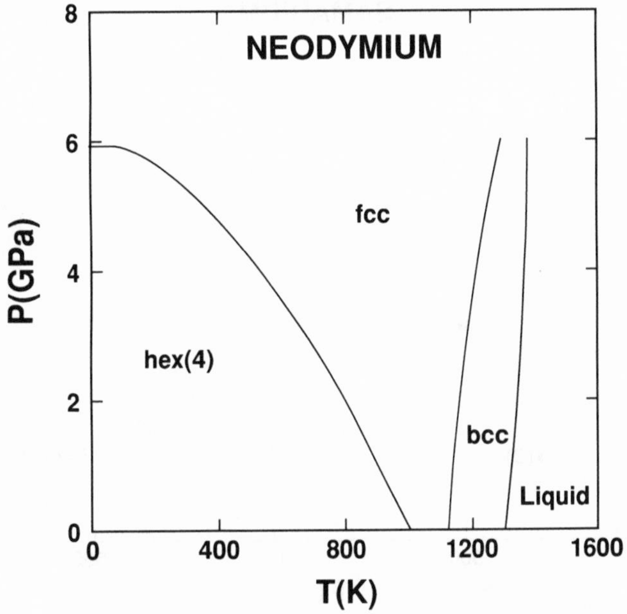

Fig. 14.6 The phase diagram of neodymium.

70. W. A. Grosshans and W. B. Holzapfel, J. Magn. Magn. Mat. **47&48**, 295 (1985).
71. J. Röhler, Physica **144B**, 27 (1986).
72. F. P. Bundy and K. J. Dunn, Phys. Rev. B **24**, 4136 (1981).
73. A. Jayaraman, Phys. Rev. **135**, A1056 (1964).
74. H. L. Skriver, Phys. Rev. Lett. **49**, 1768 (1982).
75. B. I. Min, H. J. F. Jansen, T. Oguchi, and A. J. Freeman, J. Magn. Magn. Mat. **59**, 277 (1986).
76. J. Akella, G. S. Smith, and A. P. Jephcoat, J. Phys. Chem. Solids **49**, 573 (1988).
77. N. Hamaya, K. Fuchizaki, and T. Kikegawa, High Press. Res. **4**, 375 (1990).
78. J. Staun Olsen, S. Steenstrup, and L. Gerward, Phys. Lett. **109A**, 235 (1985).
79. D. R. Stephens, J. Phys. Chem. Solids **26**, 943 (1965).
80. T. G. Ramesh, V. Shubha, and S. Ramaseshan, J. Phys. F **7**, 981 (1977).
81. M. Rieux and D. Jerome, Solid State Comm. **9**, 1179 (1971).
82. P. C. Souers and G. Jura, Science **140**, 481 (1963).
83. M. Mohan, C. Divakar, and A. K. Singh, Physica **139&140B**, 253 (1986).
84. D. B. McWhan, T. M. Rice, and P. H. Schmidt, Phys. Rev. **177**, 1063 (1969).
85. K. Syassen, G. Wortmann, J. Feldhaus, K. H. Frank, and G. Kaindl, Phys. Rev. B **26**, 4745 (1982).
86. J. F. Herbst and J. W. Wilkins, Phys. Rev. B **29**, 5992 (1984).
87. B. I. Min, T. Oguchi, H. J. F. Jansen, and A. J. Freeman, Phys. Rev. B **34**, 654 (1986).
88. J. C. Duthie and D. G. Pettifor, Phys. Rev. Lett. **38**, 564 (1977).
89. C. H. Hodges, Acta Met. **15**, 1787 (1967).
90. R. Bruinsma and A. Zangwill, Phys. Rev. Lett. **55**, 214 (1985).
91. K. A. Gschneidner, Jr., J. Less-Common Met. **114**, 29 (1985).

CHAPTER 15
The Actinides

15.1 Introduction

The actinides are a long series of radioactive metals which represent the filling of the 5*f* electron shell. The importance of nuclear power production and nuclear weaponry in the modern world guarantees a steady research effort on the physics and chemistry of the actinide metals. The rapidly decreasing half-lives of the heavier actinides have made Es the last element at present which can be studied in pure solid form. However, the chemical behavior of the heavier actinides has been studied by high-speed techniques, and a consistent picture has emerged for the systematics of the whole series. In addition, recent high-pressure DAC studies on the actinides up to Cf have revealed new patterns of phases. This experimental work together with fully relativistic band-structure calculations have clarified the relationship between the lanthanide and actinide series.

15.2 Actinium

At RTP, Ac is fcc[1]. There is no evidence for other phases at RP. Because of the intense radioactivity of Ac, high pressure XRD work has not been performed. Ac melts at 1324 K[2]. The melting curve has not been measured.

Relativistic APW calculations on Ac indicate that the 5*f* band is above the Fermi level at RP and that it has only a small effect on the properties of the solid. At higher pressure the *s-d* electron transfer is predicted to occur, as expected. Thus Ac is likely to behave much like La under pressure[3,4].

15.3 Thorium

At RTP, Th is fcc. RT compression to 100 GPa shows no change of phase[5,6]. Th transforms to bcc at 1633 K and RP[2]. The volume change is zero to within the accuracy of the measurements[1]. The fcc-bcc phase boundary has not been determined.

Th melts at 2028 K. The melting curve has not been determined.

LMTO calculations correctly predict that fcc is the stable phase of Th, and they also indicate that in Th the 5*f* band is nearly empty and that Th behaves like a tetravalent transition metal[3,7,8]. The theoretical 0 K isotherm to 40 GPa is in accurate agreement with experiment[9].

15.4 Protactinium

At RTP, Pa is ct(2), with a c/a ratio of 0.82. This structure may be considered as a distorted bcc lattice[1]. RT compression to 53 GPa shows no change of phase[10].

When Pa is quenched from high temperature, it is recovered with the fcc structure, which may be the high-temperature form[3]. The transition temperature between ct(2) and fcc at RP has not been determined, but a rough estimate is 1500 K[11].

Pa melts at 1845 K. The melting curve has not been determined.

LMTO calculations show that Pa is the first actinide with 5*f* participation in bonding[3,7]. Total-energy calculations for the tetragonal symmetry show three minima at different c/a values[8]. One of these is near the observed c/a value, although this is not the lowest minimum. More accurate calculations are expected to give better agreement with experiment, but it is evident that the 5*f* electrons are stabilizing the ct(2) structure.

15.5 Uranium

At RTP, α-U is eco(4). Compression of α-U at RT shows no change of structure to 100 GPa[12,13]. At RP and 940 K, there is a transformation to β-U, which is st(30)[1]. The precise space group of β-U is still uncertain. At 1050 K, β-U transforms to γ-U which is bcc. The α-β and β-γ phase boundaries have been measured to 3.0 GPa, where they come together in a triple point[14]. The α-γ phase boundary has been followed to 4.0 GPa[14].

At RP and 43 K, α-U exhibits a number of anomalies which suggest a phase transition[15]. XRD experiments show no change of structure, but neutron diffraction shows multiple reflections, indicating a lattice distortion. There is now broad agreement that below 43 K, α-U is in an incommensurate charge-density-wave (CDW) state, in which the atoms are displaced due to a sinusoidal modulation of the charge density[16]. Two further transitions at 37 K and 23 K may be due to different CDW symmetries. The CDW disappears above 0.6 GPa[17]. The details of the electronic structure responsible for this phenomenon are not yet determined.

The melting curve of U has been determined to 4.0 GPa[18]. The phase diagram of U is shown in Fig. 15.1.

Detailed band-structure calculations for α-U are difficult because the rather open crystal structure requires large Madelung energy corrections to the atomic-sphere approximation[8]. Total-energy comparisons of different lattices have not yet been made.

15.6 Neptunium

At RTP, α-Np is so(8), with 2 distinct types of atomic sites[1]. At 551 K, there is a transition to β-Np, st(4), also with 2 types of atomic sites. At 843 K, β-Np becomes γ-Np, which is bcc. The α and β phases may be considered as distorted bcc structures[1]. RT compression of α-Np to 52 GPa shows no phase transition[19]. However, there is a tendency toward tetragonal symmetry, which may indicate a transition at higher pressure[13].

The α-β, β-γ, and melting curves have been determined to 3.5 GPa[20]. There is a β-γ-liquid triple point at 3.2 GPa. The phase diagram of Np is shown in Fig. 15.2.

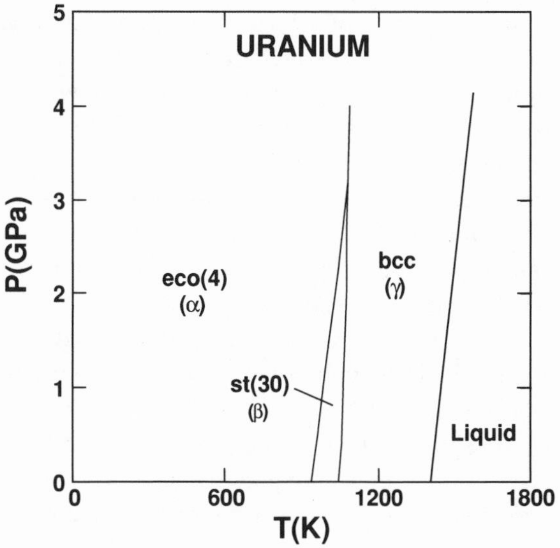

Fig. 15.1 The phase diagram of uranium.

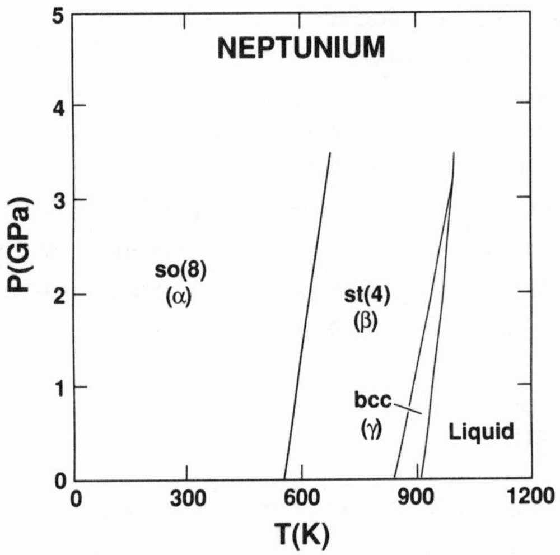

Fig. 15.2 The phase diagram of neptunium.

15.7 Plutonium

Pu has an unusually complex phase diagram, with 8 known solid phases. The low-temperature α phase is monoclinic sm(16), with 8 crystallographically distinct atomic sites[1]. This phase has been compressed at RT to 60 GPa[6,21]. One investigation finds a phase transition at ca. 40 GPa to an orthorhombic structure, possibly so(4)[21], while the other reports a high-pressure hcp phase[6]. Heating of α-Pu at RP leads to the β phase, which is also monoclinic, cm(34), with 7 distinct atomic sites[1]. Both the α and β structures are quite irregular and difficult to correlate with close-packed lattices. The α-β boundary has been measured to 14 GPa and shows a temperature maximum at ca. 6.0 GPa[22].

With increasing temperature at RP, β-Pu transforms to γ-Pu, fco(8). Below 1 GPa, γ-Pu transforms to ζ-Pu with a complex undetermined structure[23]. The ζ-Pu phase disappears in a β-ζ-liquid triple point at 2.7 GPa. Further heating at RP leads to δ-Pu, which is fcc, δ′-Pu, which is ct(2), and ε-Pu, which is bcc. All of these phases disappear at modest pressures[22,24,25]. The large volume changes accompanying the transitions at RP are shown in Fig. 15.3.

The melting curve of Pu shows a negative slope and a temperature minimum in the ε- and ζ-phase region. Above 2.7 GPa the β-liquid boundary has a positive slope. This curve has been measured to 14 GPa[22]. The phase diagram of Pu is shown in Fig. 15.4 for low pressures and in Fig. 15.5 for high pressures.

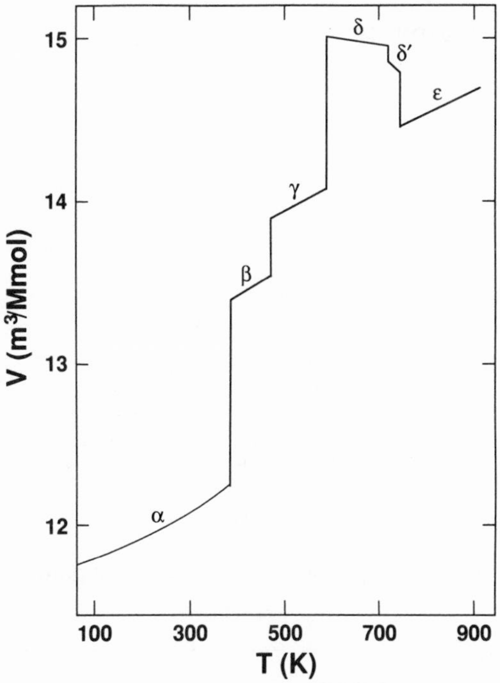

Fig. 15.3 The molar volume of plutonium at RP as a function of temperature. (From Donohue[1]. © 1974 John Wiley & Sons, Inc. Redrawn with permission.)

No theoretical structural calculations have yet been reported for Pu. The complex sequence of structures and large volume changes found in Pu suggest that major changes in electronic configuration occur with changing temperature and pressure. Alloy volume measurements suggest that the phase changes at low pressure correspond to valence changes and the stepwise localization of the 5*f* electrons[26,27].

15.8 Americium

At RTP, Am is dhcp or hex(4). Although a number of RT DAC studies have been made, there is still disagreement over the correct sequence of phases and their transition pressures[28–32]. Most agree that at about 5 GPa, Am transforms to fcc. The latest study claims a transition to an hex(6) "distorted fcc" phase at 13.5 GPa, followed by a transition to the eco(4) α-U phase at 23 GPa[32]. The 23 GPa transition has a volume change of about 6%. No further transitions are found up to 52 GPa[32].

At RP, Am shows an hex(4)-fcc transition at 1044 K, a transition to another solid phase, probably bcc, at 1347 K, and melting at 1449 K[33]. The

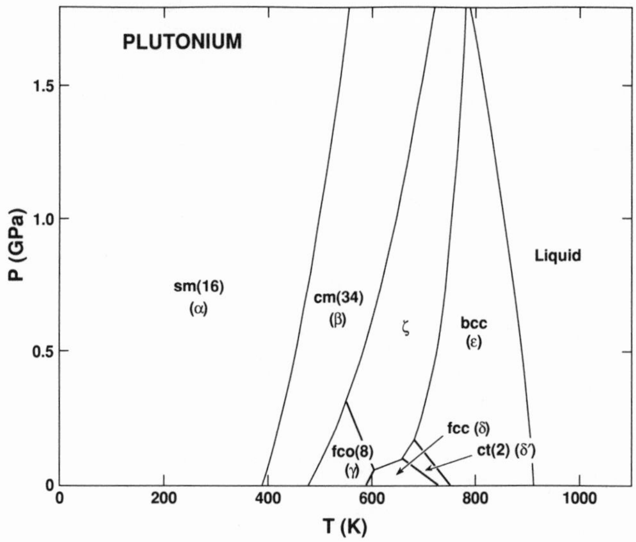

Fig. 15.4 The low-pressure phase diagram of plutonium.

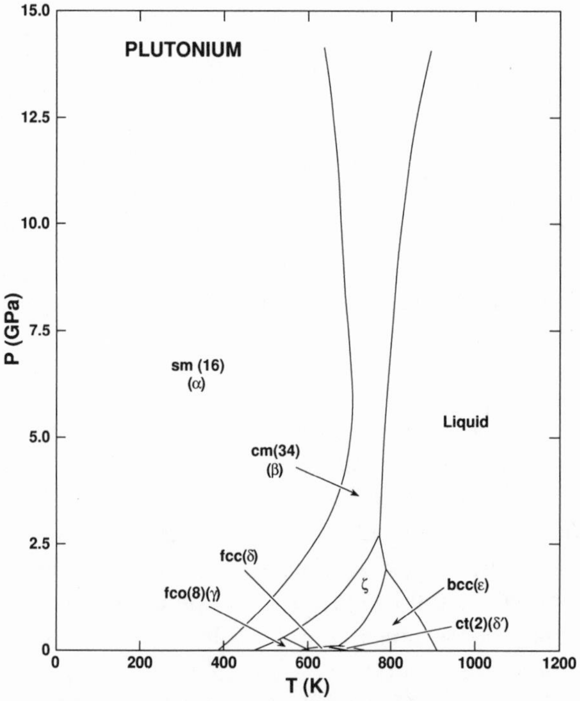

Fig. 15.5 The high-pressure phase diagram of plutonium.

hex(4)-fcc phase boundary has not been measured, but can be sketched in on the basis of its end points. The fcc-bcc and melting curves have been measured to 3.0 GPa[34]. The phase diagram of Am is shown in Fig. 15.6.

The similarity between the Am phase diagram and those of La, Pr, and Nd, is evident. This suggests that Am is the first lanthanide-like actinide, which is the result of the localization of the 5*f* electrons. This is confirmed by extensive photoemission spectroscopy, which shows that the 5*f* electrons are localized[35].

Band-structure calculations also confirm 5*f* localization. Am is the first actinide for which the LMTO spin-polarized solution with a narrow 5*f* band width is stable at RP[3,36]. Relativistic LMTO calculations correctly predict the hex(4)-fcc transition[37]. LMTO calculations also indicate a pressure-induced delocalization at ca. 10 GPa accompanied by a large volume change[3]. This volume change is not seen experimentally, but the participation of 5*f* electrons in bonding and therefore 5*f* delocalization is suggested by the appearance of the α-U structure at 23 GPa[32].

15.9 Curium

At RTP, Cm is hex(4). Under compression at RT, there is a transition to fcc at 23 GPa and another transition to eco(4) α-U at 43 GPa[38]. The fcc-eco(4)

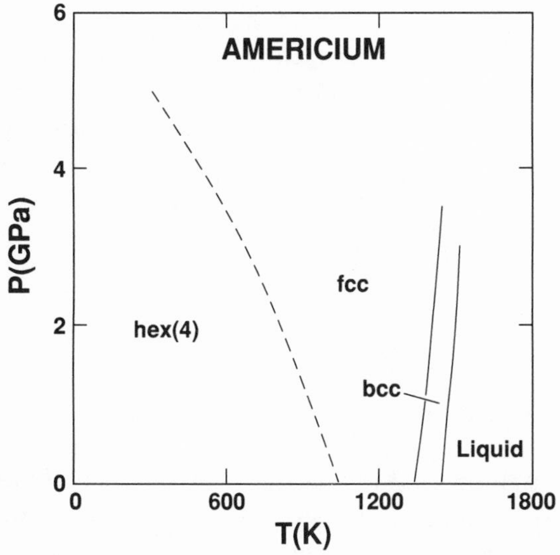

Fig. 15.6 The phase diagram of americium.

transition has a volume change of ca. 21%. No further transitions are observed up to 52 GPa[38].

The fcc phase appears at RP and ca. 1550 K[2]. There is no evidence of a bcc phase. Cm melts at 1618 K[2]. The melting curve has not been measured.

Since Cm has a half-filled 5*f* shell, it has the largest energy gain from 5*f* localization, with the result that in the sequence of transition pressures for Am, Cm, Bk, and Cf, Cm is "out of order"[39]. The very large fcc-eco(4) volume change in Cm is undoubtedly due to the 5*f* delocalization. Relativistic LMTO calculations correctly predict the hex(4)-fcc transition[37].

15.10 Berkelium

At RTP, Bk is hex(4). RT compression shows a transition to fcc at ca. 8 GPa[40,41]. A further transition to eco(4) α-U occurs at ca. 25 GPa with a 12% volume change. No further change in structure is observed to 57 GPa[40,41].

The fcc phase occurs at ca. 1250 K and RP[2]. Bk melts at 1323 K[2]. The melting curve has not been determined.

Relativistic LMTO calculations correctly predict the hex(4)-fcc transition[37].

15.11 Californium

Cf at RTP is hex(4). Compression at RT leads to fcc at ca. 17 GPa[41]. Above 30 GPa there is evidence of an hex(6) phase. At 41 GPa, Cf transforms to the eco(4) α-U structure with a 16% volume decrease. No further changes are seen up to 48 GPa[41].

An hex(4)-fcc transition is found at RP at ca. 863 K[2]. Cf melts at 1173 K[2]. The melting curve has not been measured.

Relativistic LMTO calculations correctly predict the hex(4)-fcc transition[37].

15.12 Einsteinium

Because of its high specific radioactivity and short half-life, pure Es is difficult to prepare and to work with. An electron-diffraction study has indicated an fcc structure for the metal at RTP[42]. To date, no high-pressure work has been reported.

Es melts at 1133 K at RP[2]. The melting curve has not been measured.

The large molar volume and low cohesive energy indicate that Es is divalent, like Eu and Yb among the lanthanides[2]. This correlates with the predicted $5f$-orbital contraction and the stability of the $5f^{11}7s^2$ configuration over that of $5f^{10}7s^2 6d^1$[3].

15.13 Fermium

Metallic Fm has been prepared as an alloy with lanthanide metals, and its vapor pressure has been measured[43]. From this a cohesive energy of 143 kJ/mol is obtained, which indicates a divalent state for Fm metal. The pure solid metal has not yet been prepared, but it is expected to resemble Es closely.

15.14 The Heavy Actinides and Transactinides

The remaining actinides, mendelevium (Md, element 101), nobelium (No, element 102), and lawrencium (Lr, element 103), have very short half-lives, and have not yet been prepared in pure metallic form. However, "few-atom" experiments have been carried out with these elements in order to determine their chemistry[44,45]. The results of high-speed volatility studies indicate that Md and No show increasing divalency. This is confirmed by approximate theoretical calculations, which predict a rapidly falling $5f$ band energy and stability for the $5f^n 7s^2$ electronic configuration[3]. Lr, with its complete $5f$ shell, is predicted to be trivalent, similar to Lu, and it is indeed found to show trivalent behavior[37,44,45].

The elements 104 through 109 have been prepared in high-energy nuclear-collision experiments, but chemical experiments have been performed only on 104 and 105[44,45]. This work shows that 104 and 105 form the beginning of the $6d$ transition-metal series, and resemble Hf and Ta, respectively.

15.15 Discussion

Because of the difficulty of working with the radioactive actinide elements, it is only in recent years that a comprehensive picture of actinide phase behavior has been obtained. It is now clear that the actinides differ in significant ways from the lanthanides.

The actinide molar volumes are plotted in Fig. 15.7. Unlike the 4*f* electrons in the lanthanides which participate in bonding at RP only in Ce, the 5*f* electrons are itinerant and participate in bonding in Pa, U, Np, and Pu. This is the result of the increased spatial extension of the 5*f* 's due to their orthogonality to the 4*f* 's and to relativistic effects. Thus the volumes of the early actinides form a parabola-like curve similar to those found for the transition metals, where the broad-band *d*-electrons are strongly bonding.

At Am, the stability gained by pairing the electrons in a narrow band finally exceeds the stability gained by delocalization, and the 5*f* electrons cease to participate in bonding[39,46]. There is a corresponding increase in volume. The result is a Mott-like transition which occurs "between" Pu and Am. LMTO calculations show that the spin-polarized (localized) energy becomes lower than the unpolarized band energy at Am, and that Am can be driven into a delocalized 5*f* state by pressure[3,36]. Similarly, it appears possible that Np and Pu can be driven into delocalized states by thermal expansion[47].

It is thus evident that Pu, being the last 5*f*-bonded actinide, rather closely resembles Ce, with its large α-γ volume change and anomalous melting curve. Similarly, Am rather closely resembles the lighter lanthanides Pr and Nd. The following actinides Cm, Bk, and Cf also resemble the lighter lanthanides in their phase diagrams and their slowly varying volumes. Cf shows a strong tendency toward divalency, similar to Sm, although it does not exhibit the hex(9) Sm-type structure. For the following elements the 5*f*

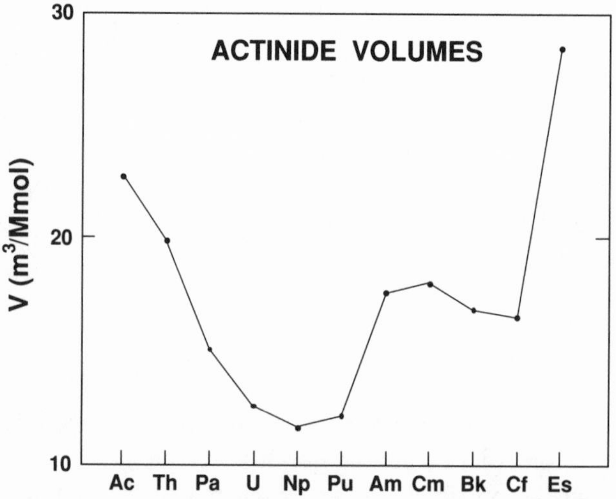

Fig. 15.7 The molar volumes of the actinides at RTP.

band energy falls more rapidly than the 4*f* band in the lanthanides, and the metallic states of Es, Fm, Md, and No are divalent, having a large increase in volume, as in Eu and Yb. Spin-polarized LMTO calculations of the zero-pressure volumes of the heavy actinides correctly reflect this progression, and the addition of spin-orbit corrections makes agreement between theory and experiment nearly quantitative[3]. Lr, with its filled 5*f* shell, is trivalent, like Lu. This suggests the lanthanide-actinide correlation shown in Fig. 15.8[2,48].

As with the lanthanides, the systematic behavior of the actinides may be illustrated in several ways:

1) The RT transition pressures are listed in Fig. 15.9. The progression of transition metal, 5*f*-bonded, and lanthanide-like structures with increasing atomic number is evident, as is the transformation of the actinide structures into 5*f*-bonded (α-U) structures at high pressure. In the sequence Am-Cm-Bk-Cf, the Cm transition transition pressures are out of order because the 5*f* localization energy reaches a maximum value for Cm[39].
2) A schematic RP binary alloy diagram similar to Fig. 14.20 may be constructed[49]. This is shown in Fig. 15.10. Of special interest are the the exotic structures and low melting points in the region between delocalized and localized 5*f* states. This feature is seen in the lanthanides only incipiently at Ce. However, under pressure, the 4*f* delocalization with corresponding low-symmetry structures is found in the early lanthanides, and the hex(4) phase is found in the heavier lanthanides, so a phase diagram similar to Fig. 15.10 might be seen in the lanthanides at very high pressure.
3) A generalized phase diagram of the kind found for the trivalent lanthanides (Fig. 14.21) is not possible because of the rapidly changing character of the 5*f* electrons[50].

Theoretical total-energy calculations of actinide structures are still difficult because the 5*f* bands are relatively narrow, because the calculations

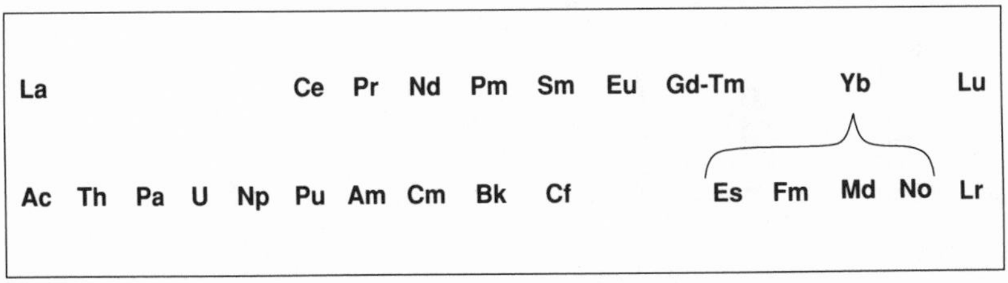

Fig. 15.8 Correlation of lanthanide and actinide elements.

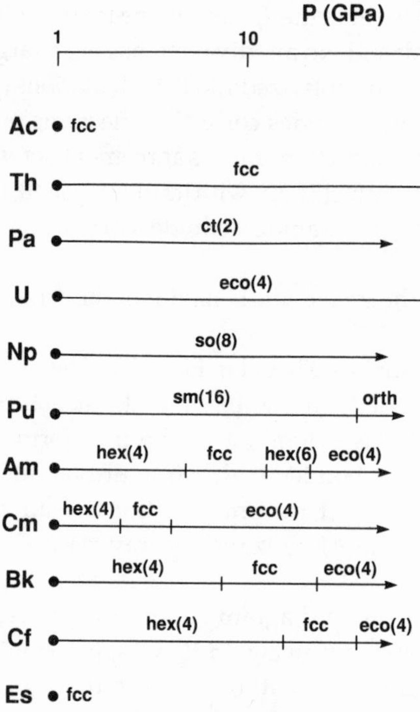

Fig. 15.9 Actinide structures at RT.

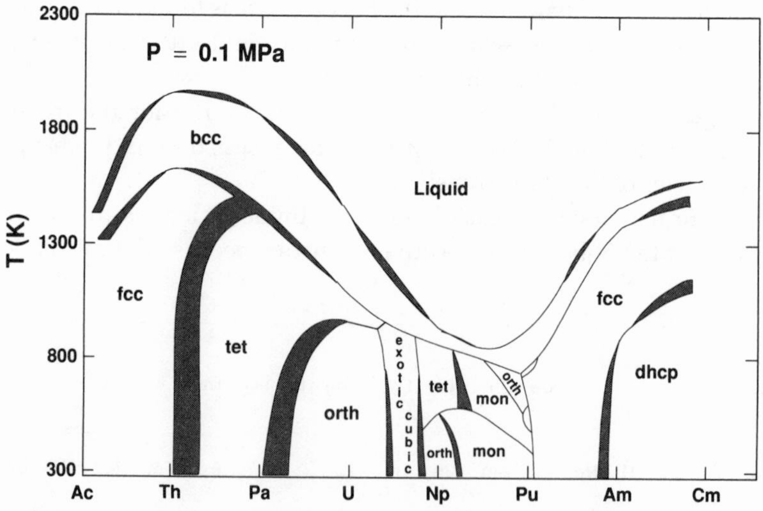

Fig. 15.10 Schematic actinide alloy phase diagram at RP.
(From Smith and Kmetko[49]. Redrawn with permission.)

must be fully relativistic, and because the atomic-sphere approximation is inappropriate for low-symmetry phases such as α-U and α-Pu. Nevertheless, approximate calculations correctly indicate the general trend of the actinide electronic structures, and a reasonably good theoretical understanding of the actinide group has emerged.

For Further Reading

A. J. Freeman and G. H. Lander, eds., *Handbook on the Physics and Chemistry of the Actinides* (North-Holland, Amsterdam, 1984) vol. 1, and further volumes in the series.

J. J. Katz, G. T. Seaborg, and L. R. Morss, eds., *The Chemistry of the Actinide Elements*, 2d ed. (Chapman and Hall, London, 1986) 2 vols.

U. Benedict, "Structural Data of the Actinide Elements and of Their Binary Compounds with Non-metallic Elements," J. Less-Common Met. **128**, 7 (1987).

References

1. J. Donohue, *The Structures of the Elements* (Wiley, New York, 1974) chap. 5.
2. J. W. Ward, P. D. Kleinschmidt, and D. E. Peterson, in *Handbook on the Physics and Chemistry of the Actinides*, A. J. Freeman and C. Keller, eds. (North-Holland, Amsterdam, 1986) vol. 4, chap. 7.
3. M. S. S. Brooks, B. Johansson, and H. L. Skriver, in *Handbook on the Physics and Chemistry of the Actinides*, A. J. Freeman and G. H. Lander, eds. (North-Holland, Amsterdam, 1984) vol. 1, chap. 3.
4. M. Dakshinamoorthy and K. Iyakutti, Phys. Rev. B **30**, 6943 (1984).
5. G. Bellussi, U. Benedict, and W. B. Holzapfel, J. Less-Common Met. **78**, 147 (1981).
6. J. Akella, Q. Johnson, G. S. Smith, and L. C. Ming, High Press. Res. **1**, 91 (1988).
7. H. L. Skriver, O. K. Andersen, and B. Johansson, Phys. Rev. Lett. **41**, 42 (1978).
8. H. L. Skriver, Phys. Rev. B **31**, 1909 (1985).
9. H. L. Skriver and J.-P. Jan, Phys. Rev. B **21**, 1489 (1980).
10. U. Benedict, J. C. Spirlet, C. Dufour, I. Birkel, W. B. Holzapfel, and J. R. Peterson, J. Magn. Magn. Mat. **29**, 287 (1982).
11. J. C. Spirlet, E. Bednarczyk, and W. Muller, J. Less-Common Met. **92**, L27 (1983).
12. J. Akella, G. Smith, and H. Weed, J. Phys. Chem. Solids **46**, 399 (1985).
13. J. Akella, G. S. Smith, R. Grover, Y. Wu, and S. Martin, preprint.
14. W. Klement, Jr., A. Jayaraman, and G. C. Kennedy, Phys. Rev. **129**, 1971 (1963).
15. E. S. Fisher and H. J. McSkimin, Phys. Rev. **124**, 67 (1961).
16. G. H. Lander, J. Magn. Magn. Mat. **29**, 271 (1982).
17. H. G. Smith and G. H. Lander, Phys. Rev. B **30**, 5407 (1984).
18. J. Ganguly and G. C. Kennedy, J. Phys. Chem. Solids **34**, 2272 (1973).
19. S. Dabos, C. Dufour, U. Benedict, and M. Pagès, J. Magn. Magn. Mat. **63&64**, 661 (1987).

20. D. R. Stephens, J. Phys. Chem. Solids **27**, 1201 (1966).
21. S. Dabos-Seignon, Thesis, University of Paris VI, 1987.
22. C. Roux, P. le Roux, and M. Rapin, J. Nucl. Mat. **40**, 305 (1971).
23. J. R. Morgan, in *Plutonium 1970 and Other Actinides*, W. N. Miner, ed. (Metallurgical Soc. AIME, New York, 1970) p. 669.
24. D. R. Stephens, J. Phys. Chem. Solids **24**, 1197 (1963).
25. R. G. Liptai and R. J. Friddle, J. Less-Common Met. **10**, 292 (1966).
26. Z. Fisk, R. O. Elliot, R. E. Tate, and R. B. Roof, in *Actinides 1981 Abstracts* (Lawrence Berkeley Laboratory, Berkeley, 1981) p. 208.
27. P. Weinberger, A. M. Boring, and J. L. Smith, Phys. Rev. B **31**, 1964 (1985).
28. J. Akella, Q. Johnson, W. Thayer, and R. N. Schock, J. Less-Common Met. **68**, 95 (1979).
29. R. B. Roof, R. G. Haire, D. Schiferl, L. A. Schwalbe, E. A. Kmetko, and J. L. Smith, Science **207**, 1353 (1980).
30. G. S. Smith, J. Akella, R. Reichlin, Q. Johnson, R. N. Schock, and M. Schwab, in Ref. 26, p. 218.
31. R. B. Roof, in Ref. 26, p. 213.
32. U. Benedict, J. P. Itié, C. Dufour, S. Dabos, and J. C. Spirlet, Physica **139&140B**, 284 (1986).
33. A. G. Seleznev, N. S. Kosulin, V. M. Kosenkov, V. D. Shushakov, V. A. Stupin, and V. A. Demeshkin, Fiz. Metal. Metalloved. **44**, 654 (1977) [Phys. Met. Metallog. **44**, 180 (1977)].
34. D. R. Stephens, H. D. Stromberg, and E. M. Lilley, J. Phys. Chem. Solids **29**, 815 (1968).
35. J. R. Naegele, L. Manes, J. C. Spirlet, and W. Muller, Phys. Rev. Lett. **52**, 1834 (1984).
36. H. L. Skriver, O. K. Andersen, and B. Johansson, Phys. Rev. Lett. **44**, 1230 (1980).
37. O. Eriksson, M. S. S. Brooks, and B. Johansson, preprint.
38. U. Benedict, R. G. Haire, J. R. Peterson, and J. P. Itié, J. Phys. F **15**, L29 (1985).
39. B. Johansson, H. L. Skriver, N. Mårtensson, O. K. Andersen, and D. Glötzel, Physica **102B**, 12 (1980).
40. R. G. Haire, J. R. Peterson, U. Benedict, and C. Dufour, J. Less-Common Met. **102**, 119 (1984).
41. U. Benedict, J. R. Peterson, R. G. Haire, and C. Dufour, J. Phys. F **14**, L43 (1984).
42. R. G. Haire and R. D. Baybarz, J. Phys. (Paris) **40**, C4-101 (1979).
43. R. G. Haire and J. K. Gibson, J. Chem. Phys. **91**, 7085 (1989).
44. E. K. Hulet, Radiochim. Acta **32**, 7 (1983).
45. R. J. Silva, in *The Chemistry of the Actinide Elements*, 2d ed., J. J. Katz, G. T. Seaborg, and L. R. Morss, eds. (Chapman and Hall, London, 1986) chap. 13.
46. B. I. Min, H. J. F. Jansen, T. Oguchi, and A. J. Freeman, J. Magn. Magn. Mat. **61**, 139 (1986).
47. Y. K. Vohra and W. B. Holzapfel, Phys. Lett. **89A**, 149 (1982).
48. B. Johansson and A. Rosengren, Phys. Rev. B **11**, 2836 (1975).
49. J. L. Smith and E. A. Kmetko, J. Less-Common Met. **90**, 83 (1983).
50. U. Benedict, W. A. Grosshans, and W. B. Holzapfel, Physica **144B**, 14 (1986).

CHAPTER 16
The Liquid-Vapor Transition

16.1 Introduction

In addition to the solid-solid and melting transitions discussed in this book, every element also has a liquid-vapor (LV) transition which ends in a critical point. The LV pressures are relatively low, and on most of the phase diagrams which have been shown, the vaporization curve would be invisible. Nevertheless, the universality of this phase behavior and the special character of the critical point make a general discussion worthwhile.

16.2 Experimental

In the experimental study of the LV transition, the usual measured variables are pressure (P), temperature (T), vapor (or gas) volume (V_g), and liquid volume (V_l). The density $\rho = 1/V$ is also a useful variable. In the P-T plane, the vaporization curve $P(T)$ is a smoothly increasing function of T, which ends in a critical point. For every element except He, the vaporization curve also ends in a solid-liquid-vapor triple point at low T. In the case of He, the solid does not exist at low pressure, and there is no triple point. In the T-ρ plane, the two-phase LV mixture occurs within a dome-shaped region defined by the ρ_g and ρ_l curves. The P-T and T-ρ curves are sketched in Fig. 16.1

The P-T trajectory of the LV transition is described by the Clausius-Clapeyron equation:

$$\frac{dP}{dT} = \frac{\Delta S}{\Delta V} = \frac{\Delta H}{T\Delta V}. \qquad (16.1)$$

At low temperatures, $V_g >> V_l$ and the ideal gas law holds:

$$\Delta V = V_g - V_l \cong V_g = \frac{RT}{P}. \qquad (16.2)$$

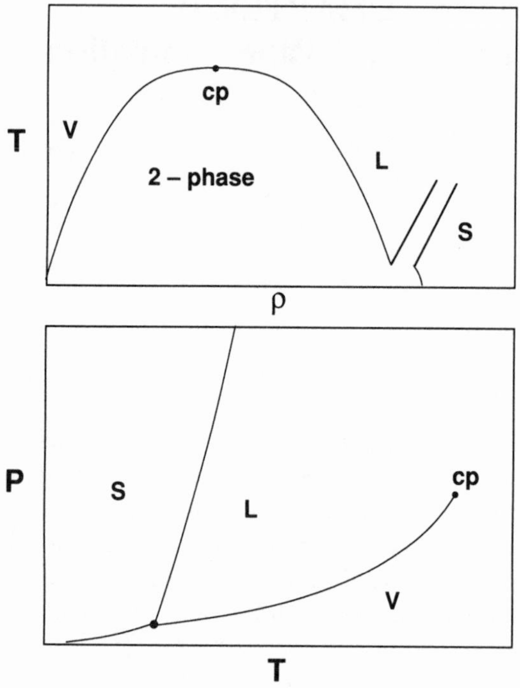

Fig. 16.1 General temperature-density and pressure-temperature liquid-vapor phase diagrams.

Then

$$\ln P \;=\; -\,\frac{\Delta H}{RT} \;+\; \text{const}, \tag{16.3}$$

where ΔH is a constant, equal to the cohesive energy or enthalpy of vaporization at $T = 0$ K. Eq (16.3) is strictly valid only at low temperatures. However, the errors in the assumptions for ΔH and ΔV rather accurately cancel, and the prediction of Eq. (16.3) holds remarkably well all the way to the critical point[1]. This is illustrated in Fig. 16.2 for three different elements, Ar[2], Cs[3], and Hg[4].

Another useful fact is the "law of rectilinear diameter," which states that the average $(\rho_g + \rho_l)/2$ vs. T is linear from triple point to critical point. This is known to be untrue very close to the critical point, but it is accurately obeyed by simple molecular elements over a broad range of temperatures[1]. Noticeable deviations are seen in metals and in dissociating molecular elements. The T-ρ coexistence curves in reduced coordinates are shown for three elements in Fig. 16.3[2–4].

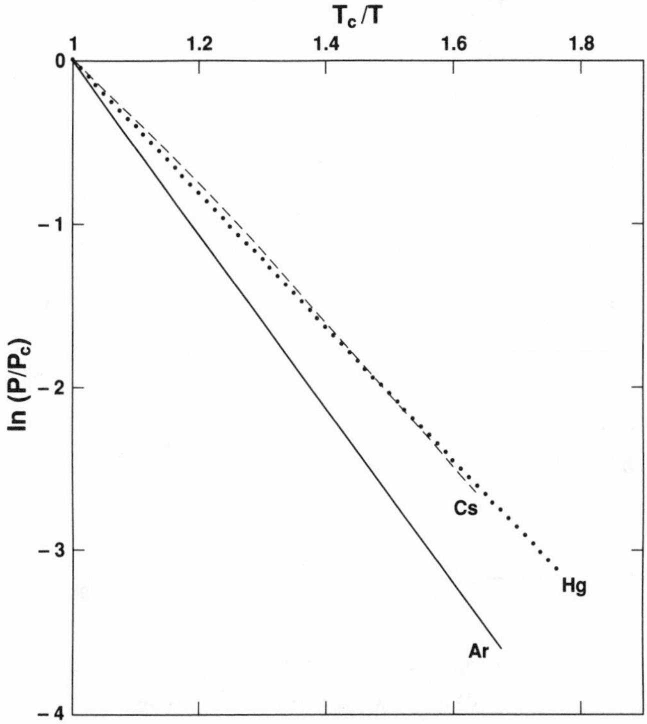

Fig. 16.2 Comparison of $\ln(P/P_c)$ vs. T_c/T for argon, cesium, and mercury.

If the principle of corresponding states were obeyed by Ar, Cs, and Hg, then the curves shown in Figs. 16.2 and 16.3 would agree accurately with one another. That they do not implies that corresponding states does not hold among all of the elements, which is well known. Among certain groups of elements, however, the rule is accurately obeyed. For the simple molecular elements Ar, Kr, Xe, N_2, and O_2, corresponding states holds accurately[5]. H_2 and He do not correspond with the heavier gases because of quantum effects. Rb and Cs correspond well with each other[6], but not with Hg.

The critical-point singularity has received intense study, both experimentally and theoretically. The critical singularity is defined in terms of exponents such as α, β, γ, and δ:

$$C_v \propto |T - T_c|^{-\alpha} , \tag{6.4}$$

$$|\rho - \rho_c| \propto |T - T_c|^{\beta} , \tag{6.5}$$

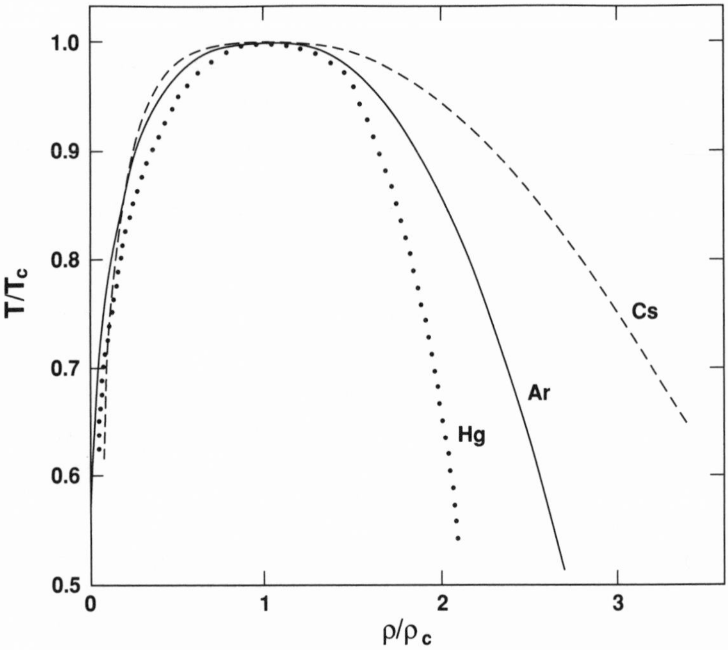

Fig. 16.3 Comparison of T/T_c vs. ρ/ρ_c coexistence curves for argon, cesium, and mercury.

$$-\frac{\partial P}{\partial V} \propto |T - T_c|^{\gamma}, \tag{6.6}$$

$$|P - P_c| \propto |\rho - \rho_c|^{\delta}. \tag{6.7}$$

It is difficult to obtain these exponents for liquid metals because of the high temperatures and pressures required, but recent experimental work strongly implies that the exponents for metals are the same as for insulators. The measured exponents are summarized[3,7–9] in Table 16.1.

The experimentally determined critical points of the elements are tabulated[3,6,8,10–16] in Table 16.2. From these data we can see that the critical density typically lies in the range 1/2 to 1/5 of the normal solid density and that the critical temperature scales approximately with the cohesive energy. These observations make it possible to estimate the critical points of elements where measurements are not yet possible. For nearly all of the metallic elements, the critical temperatures will be beyond the reach of any static experiment because they exceed the melting point of any possible container material. For these elements the measurement of the critical point must be a dynamic experiment, such as the isobaric-heating

TABLE 16.1 Critical Exponents

	α	β	γ	δ	Ref.
Xe	0.08	0.344	1.203	4.4	7
Cs	0.13 ± 0.03	0.355 ± 0.01	—	—	3
Hg	—	0.32	1.187	4.71	7
S	—	0.33 ± 0.01	1.16 ± 0.2	4.72 ± 0.1	8
van der Waals	0	0.5	1.0	3.0	9
3-D Ising	0.125	0.313	1.25	5.0	9

TABLE 16.2 Critical Points of the Elements

Element	T_c (K)	P_c (MPa)	V_c (m^3/Mmol)	Z_c	V_c/V_o	RT_c/E_{coh}	Ref.
H_2	33.19	1.315	64.94	0.309	2.81	0.379	10
D_2	38.34	1.665	59.52	0.311	2.98	0.280	10
^3He	3.310	0.1147	72.5	0.302	1.97	1.338	11
^4He	5.190	0.2275	57.54	0.303	2.09	0.724	11
N_2	126.2	3.39	89.5	0.289	3.28	0.152	12
O_2	154.6	5.043	73.4	0.288	3.53	0.148	12
F_2	144.3	5.215	66.2	0.288	3.42	0.130	12
Ne	44.40	2.653	41.8	0.300	3.12	0.197	11
Na	2485	25.55	76.67	0.095	3.38	0.192	13
P	994	—	—	—	—	0.026	12
S	1313	20.3	55.3	0.103	—	0.396	8
Cl_2	417	7.70	124.7	0.275	3.59	0.115	12
Ar	150.70	4.86	74.9	0.290	3.32	0.162	11
K	2198	15.5	—	—	—	0.203	13
As	1733	9.87	27.78	0.0190	2.14	0.048	14
Se	1903	38.0	42.7	0.103	2.60	0.077	15
Br_2	584	10.3	127	0.269	3.22	0.100	12
Kr	209.5	5.52	91.3	0.289	3.38	0.156	11
Rb	2017	12.45	293	0.218	5.27	0.198	3
I_2	819	10.3	155	0.234	3.15	0.109	12,16
Xe	289.72	5.840	118	0.286	3.40	0.150	11
Cs	1924	9.25	351	0.203	5.28	0.199	3
Hg	1750	167.3	34.77	0.400	2.50	0.227	6
Rn	378	6.3	—	—	—	—	12

technique or adiabatic release from shock compression. Both of these methods have been used on Pb to reach the critical region, and the critical point has been then estimated by theoretical models[17,18]. For Pb, $T_c \approx 5000$–6000 K, and $P_c \approx 0.2$ GPa. Much more experimental work of this type on the elements is needed.

Of great interest in Table 16.2 is the dimensionless parameter $Z_c = P_cV_c/RT_c$. This number varies over a narrow range from about 0.2 for Cs to about 0.4 for Hg. The simple gases have a Z_c of about 0.30. The number $Z_c = 0.1$ for S assumes V_c in units of m^3/(Mmol S). Since the actual molecular species at the critical point are a mixture of molecules S_n, the value of Z_c contains information about n.

The LV transition in the monatomic rare gases and the simple diatomic gases has been measured with high accuracy. The LV coexistence curve is quite symmetric and the rectilinear-diameter law holds with good accuracy. Since the band-gap energies in these fluids are much greater than the thermal energies at the critical point, it can be assumed that the interaction potential is accurately pairwise and independent of density and temperature throughout the LV transition region. Hence in these fluids the LV transition is a purely statistical-physics phenomenon, unperturbed by changing interactions.

In liquid S, the liquid at low temperatures is composed of S_8 rings. As the temperature is increased, the rings break up into chains, which in turn break up into S_n molecular fragments. At the critical point there is a distribution of molecular sizes[19]. In this case the critical point is determined by a changing chemical equilibrium as well as by the usual energy-entropy balance. The result is a less-symmetric coexistence curve.

For liquid metals, expansion leads to a metal-nonmetal transition, since the low-density atomic gas must be an insulator. The details of this transition have been difficult to observe because of the high temperatures and pressures required. In Cs, the conductivity along the critical isotherm shows a very rapid change in the vicinity of the critical density[20]. This metal-nonmetal transition is evidently not a true phase transition, although it does appear to be centered on the critical density. Because of the very different interactions in metallic and insulating states, the coexistence curve is asymmetric and the function $(\rho_g + \rho_l)/2$ is noticeably curved[3,21]. In Hg, the behavior upon expansion is quite different from that of Cs[22]. At 9 Mg/m^3, there is a metal-insulator transition well before the LV critical point at 5.8 Mg/m^3. Another transition is found at 3 Mg/m^3, between two insulating vapor states.

16.3 Theoretical

The LV transition is easily explained in terms of a liquid phase stabilized by low energy and a vapor phase stabilized by high entropy. A model intermolecular potential with a short-range repulsion and a longer-ranged attraction such as the Lennard-Jones 6–12 potential is sufficient to produce a LV transition with a critical point[23].

No one doubts that an exact calculation of the equation of state of a fluid governed by a Lennard-Jones or exponential-six potential would show all of the features of the liquid-vapor transition observed in real insulating fluids, including the correct critical-point singularities. However, such calculations are not yet possible, and approximate theories must be used. In one group of theories, based on liquid models, realistic interactions are used, and the global features of the LV transition are predicted, but the details of the critical singularity are incorrect. These models may be used to estimate vaporization curves and critical points. In another group of theories, based on lattice-gas models, the interaction-potential model is greatly simplified in order to permit accurate calculation of the critical exponents, but now correlation with real fluid thermodynamics is lost.

An example of the first group of theories is a very simple van der Waals-type model based on the hard-sphere equation of state with an attractive mean field[24]:

$$P = \frac{NkT}{V} \frac{(1 + y + y^2 - y^3)}{(1 - y)^3} - \frac{a}{V^2}, \tag{16.8a}$$

$$a = E_{coh} V_o , \tag{16.8b}$$

$$y = \frac{\pi N \sigma^3}{6V} , \tag{16.8c}$$

$$\sigma = \left(\frac{2.7 V_L}{\pi N} \right)^{1/3} . \tag{16.8d}$$

The empirical parameters are E_{coh} = cohesive energy; V_o = solid volume; and V_L = liquid volume at the triple point. The parameters a and σ are determined from these data. Solving for the critical point, $\partial P/\partial V = \partial^2 P/\partial V^2 = 0$, we get:

$$V_c = 4.014 N \sigma^3 , \tag{16.9a}$$

$$T_c = 0.7232 \frac{a}{R V_c} , \tag{16.9b}$$

$$P_c = 0.2596 \frac{a}{V_c^2} , \tag{16.9c}$$

$$Z_c = 0.3590 . \tag{16.9d}$$

These simple scaling laws clearly relate the behavior of the critical point with the empirical constants a and σ. Thus the critical volume is proportional to the atomic volume, and the critical temperature is proportional to the cohesive energy. We expect the critical constants predicted by this theory to be semiquantitative.

More realistic models must make use of accurate representations of the intermolecular potential. If this potential is known, then Monte Carlo, Molecular Dynamics, or liquid perturbation theory may be used to obtain the equation of state and the LV transition, including the critical point. Where the interaction potential is volume-dependent or is not well known, as in metals, an indirect approach is to fit experimental liquid EOS data with a simple theoretical model with adjustable parameters, and then to compute the critical point. An example of this approach is the "soft-sphere" model, which uses the inverse-power potential $\phi(r) = \varepsilon(\sigma/r)^n$ as the basis for fitting EOS data[25]. For metals with high cohesive energies such as W and Ta, the soft-sphere model predicts $T_c \approx 10000$ K and $P_c \approx 1$ GPa. These are probably the highest possible critical temperature and pressure values.

For simple molecular gases like H_2, Ar, and N_2, the interaction potential is unchanged in passing from liquid to vapor, and phenomena like dissociation and ionization do not occur to complicate the transition. Simple potentials like the exponential-six give good fits to experimental EOS data, and the critical points predicted by statistical-mechanical theories using these potentials are also in good agreement with experiment[26].

The theory of the metallic LV transition is much more complex than for molecular insulators because the metallic interactions are volume dependent, and there is a transition to a very different insulating potential. Although NFE liquid-metal theory can be combined with ideal-gas statistical mechanics to predict the coexistence curve at subcritical temperatures[27], accurate first-principles theoretical predictions near the critical point are not yet available, and current modeling has focused on the qualitative features of metals in the critical region. One-electron band-structure theory cannot represent the metal-insulator transition because of the inadequate treatment of electron correlation.

For Cs, the 6*s* band is half-filled, so expanded Cs will remain metallic until the electron-correlation energy favors a gas of independent atoms. This is indicated in Fig. 16.4. Spin-polarized LMTO[28], ionization equilibrium[29], Anderson localization[30], and percolation[31] models have been used to explain the metal-insulator transition in Cs. Semiquantitative agreement with experiment is possible.

In Hg, the dense solid and liquid are metallic because of the overlap of the 6*s* and 6*p* bands. At ca. 9 Mg/m^3, the bands separate, creating an insulating gap as shown in Fig. 16.4[32]. Liquid disorder and thermal excitation make the gap a "pseudogap," but this is apparently sufficient to drive the metal-insulator transition. There is controversy over the microstructure of the insulating fluid. One suggestion is a fluid of droplets stabilized by thermally ionized electrons[33,34]. An opposing "excitonic insulator" concept assumes that the fluid takes the form of droplets stabilized by electrons in Frenkel exciton states[35]. The transition at 3 Mg/m^3 would presumably represent the onset of a classical insulating gaseous state.

A very different group of theories deals with the critical-point singularity. The paradigm of this approach is the 3-dimensional Ising lattice gas. Here only nearest-neighbor interactions are considered. Approximating

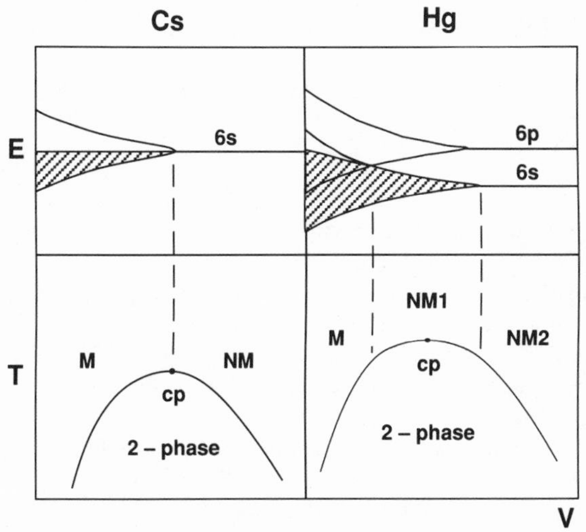

Fig. 16.4 Correlation of band structures and phase diagrams for cesium and mercury.

the lattice-gas partition function with a mean field yields a critical point with the van der Waals exponents (Table 16.1). An exact expansion of the partition function yields exponents very close to those measured experimentally. Evidently the very-long-wavelength fluctuations at the critical point destroy any dependence on the details of the interaction potential.

The Renormalization Group theory of the critical point predicts a set of exponents for each "universality class" of fluids, dependent on dimensionality D and the range of the interactions[9]. As far as is known, all of the chemical elements are in the same universality class. The critical exponents are determined by expanding about $D = 4$, where the exponents have the classical values. The predictions are in remarkable agreement with experiment[9].

16.4 Discussion

For low-boiling insulating fluids such as Ar and N_2, there is an abundance of experimental data on the LV transition and the critical point, and the theoretical understanding of these phenomena is good. For higher-boiling elements, including covalent and metallic elements, experiments are more difficult and theoretical models are less accurate. The metal-nonmetal transition and its connection with the LV transition is a major challenge both experimentally and theoretically. Given the experience so far with Hg, it would not be surprising if new exotic states of matter occurred in hot, expanded metals which are presently inaccessible to experimental work.

References

1. J. S. Rowlinson and F. L. Swinton, *Liquids and Liquid Mixtures,* 3d ed. (Butterworth Scientific, London, 1982) chaps. 2, 3.
2. N. B. Vargaftik, *Handbook of Physical Properties of Liquids and Gases,* 2d ed. (Hemisphere, Washington, 1975) pp. 543–544.
3. S. Jüngst, B. Knuth, and F. Hensel, Phys. Rev. Lett. **55**, 2160 (1985).
4. H. v. Tippelskirch, E. U. Franck, and F. Hensel, Ber. Bunsenges. Phys. Chem. **79**, 889 (1975).
5. E. A. Guggenheim, *Applications of Statistical Mechanics* (Oxford University Press, Oxford, 1966) chap. 3.
6. F. Hensel, S. Jüngst, B. Knuth, H. Uchtmann, and M. Yao, Physica **139&140B**, 90 (1986).
7. M. M. Korsunskii, Zh. Eksp. Teor. Fiz. **89**, 875 (1985) [Sov. Phys. JETP **62**, 502 (1985)].

8. R. Fischer, R. W. Schmutzler, and F. Hensel, Ber. Bunsenges. Phys. Chem. **86**, 546 (1982).
9. R. Balescu, *Equilibrium and Nonequilibrium Statistical Mechanics* (Wiley, New York, 1975) chaps. 9, 10.
10. P. C. Souers, *Hydrogen Properties for Fusion Energy* (University of California Press, Berkeley, Los Angeles, London, 1986) chap. 4.
11. R. K. Crawford, in *Rare Gas Solids*, M. L. Klein and J. A. Venables, eds. (Academic, London, 1977) vol. 2, chap. 11.
12. J. F. Mathews, Chem. Revs. **72**, 71 (1972).
13. J. Magill and R. W. Ohse, in *Handbook of Thermodynamic and Transport Properties of Alkali Metals* (Blackwell, Oxford, 1985) p. 73.
14. F. Hensel, personal communication.
15. S. Hosokawa and K. Tamura, J. Non-Cryst. Solids **117/118**, 52 (1990).
16. M. Yao, N. Nakamura, and H. Endo, Z. Phys. Chem. NF **157**, 569 (1988).
17. W. M. Hodgson, Lawrence Livermore Laboratory Report UCRL-52493, 1978.
18. V. E. Fortov and I. T. Iabukov, *Physics of Nonideal Plasma* (Hemisphere, New York, 1990) chap. 3.
19. K. Tamura, H. P. Seyer, and F. Hensel, Ber. Bunsenges. Phys. Chem. **90**, 581 (1986).
20. F. Hensel, in *Strongly Coupled Plasma Physics*, F. J. Rogers and H. E. Dewitt, eds. (Plenum, New York, 1987) p. 381.
21. V. F. Kozhevnikov, Zh. Eksp. Teor. Fiz. **97**, 541 (1990) [Sov. Phys. JETP **70**, 298 (1990)].
22. F. Hensel, Mat. Res. Soc. Symp. Proc. **22**, 3 (1984).
23. J. P. Hansen and L. Verlet, Phys. Rev. **184**, 151 (1969).
24. D. A. Young and B. J. Alder, Phys. Rev. A **3**, 364 (1971).
25. D. A. Young, Lawrence Livermore Laboratory Report. UCRL-52352, 1977.
26. D. A. McQuarrie, *Statistical Mechanics* (Harper & Row, New York, 1976) chaps. 12–14.
27. S. M. Osman and W. H. Young, Phys. Chem. Liq. **17**, 181 (1987).
28. P. J. Kelly and D. Glötzel, Phys. Rev. B **33**, 5284 (1986).
29. J. P. Hernandez, Phys. Rev. Lett. **57**, 3183 (1986).
30. J. R. Franz, Phys. Rev. B **29**, 1565 (1984).
31. T. Odagaki, N. Ogita, and H. Matsuda, J. Phys. Soc. Japan **39**, 618 (1975).
32. F. Hensel, Angew. Chem. Int. Ed. **13**, 446 (1974).
33. W. Hefner, B. Sonneborn-Schmick, and F. Hensel, Ber. Bunsenges. Phys. Chem. **86**, 844 (1982).
34. J. P. Hernandez, Phys. Rev. Lett. **48**, 1682 (1982).
35. L. A. Turkevich and M. H. Cohen, Ber. Bunsenges. Phys. Chem. **88**, 292 (1984).

CHAPTER 17
Overview

17.1 Introduction

The theme of this book has been the impressive advances in experimental and theoretical condensed-matter physics in recent years. These advances have come from many different sources and have converged in a mass of new experimental data and theoretical calculations. The result is that the behavior of the elements at high pressure is now far better understood than it was only 10 years ago.

The advent of routine measurements above 100 GPa means that even the most incompressible elements can be compressed significantly, and that phase transitions at very high pressure can be found. The new technology has been applied to very compressible elements such as H_2 and He, and to the previously unavailable heavy actinides, so there is a more complete experimental coverage of the entire periodic table as well as an overall increase in compression. The temperature dimension of the experimental work has not expanded as rapidly as the pressure dimension, but new DAC and shock-wave techniques are already beginning to rectify this.

The theoretical understanding of the patterns revealed by the experimental work is now good, although prediction of complex crystal structures from first principles is still difficult. The most powerful theoretical approach used in studying the elements at high pressures is the electron-band-structure theory in the local-density approximation. Using only the nuclear charge as input, this theory obtains accurate total energies of crystal structures and can predict phase transitions between phases at 0 K. These calculations have been outstandingly successful.

In this chapter I present an overview of our present understanding of the elements at high pressure and temperature.

17.2 Equation of State

The periodicity of the elements is readily understood in terms of the filling of atomic energy levels as a function of the atomic number Z. A deeper understanding of this periodicity at high pressures requires a knowledge of the band structure of the solid elements as a function of pressure and Z. Since this information is not readily obtainable from experiments, other experimental variables are discussed here.

The atomic volumes of the elements at $T = 0$ K and for various pressures are shown in Fig. 17.1. The $P = 0$ curve very clearly shows the periodicity of the elements and the very large differences in atomic volume among them. The sharp peaks are the alkali metals, bracketed by the rare gas and alkaline-earth elements. The three central parabolas are the transition metals. The lanthanides form a "ramp" with negative slope (the lanthanide contraction) and two anomalies, namely Eu and Yb. The actinides show a very different pattern, namely a transition-metal-like parabola followed by the beginning of a lanthanide-like "ramp," and then a large volume jump at Es. This is a clear demonstration of the differences between the lanthanides and actinides.

At 0.01 TPa, the V vs. Z plot in Fig. 17.1 shows the same periodicity, but now with a much reduced ratio of large to small volumes. The most compressible elements shrink rapidly under pressure, while the least

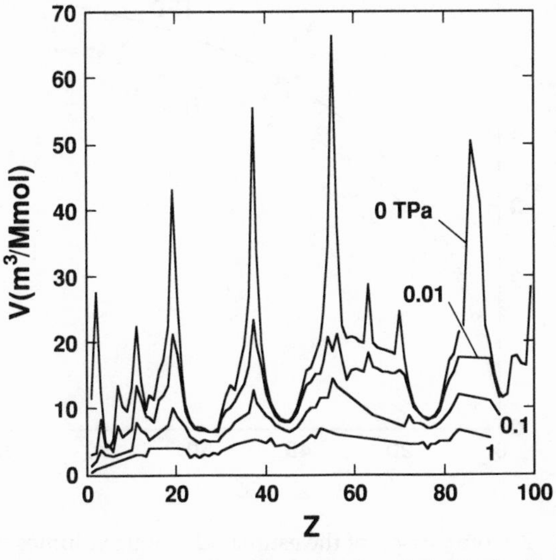

Fig. 17.1 The experimentally determined and theoretically estimated atomic volumes of the elements at 0, 0.01, 0.1, and 1 TPa.

compressible change hardly at all. At 0.1 TPa, there are still many experimental data available, but most of the volumes must now be estimated from fits to the Birch-Murnaghan equation derived from lower-pressure isotherms or shock-wave data, or from total-energy band-structure calculations. At this pressure, the convergence of the elemental volumes is far enough along to show a trend. At 1 TPa, only the least compressible element volumes can be estimated from Birch-Murnaghan fits, and most must be calculated from band-structure calculations. At this pressure, the volume curve has become much smoother, and the agreement with Thomas-Fermi calculations shown in Fig. 17.2 shows that the details of band structure which make the elements different from one another at low pressure have been largely eliminated[1].

As I have described in the body of this book, the best band-structure calculations do very well in predicting the $P = 0$ volumes of the elements. In general, these same theories will do just as well for higher pressures. But as the band structure is simplified by the breakdown of covalent bonds and the approach to close-packed lattices at high pressure, simpler theories such as the Thomas-Fermi-Dirac also begin to describe the experimental data. At extreme pressure, the electrons are completely free, and all of the theories lead to the same result. The "scale pressure" for the reduction of volumes

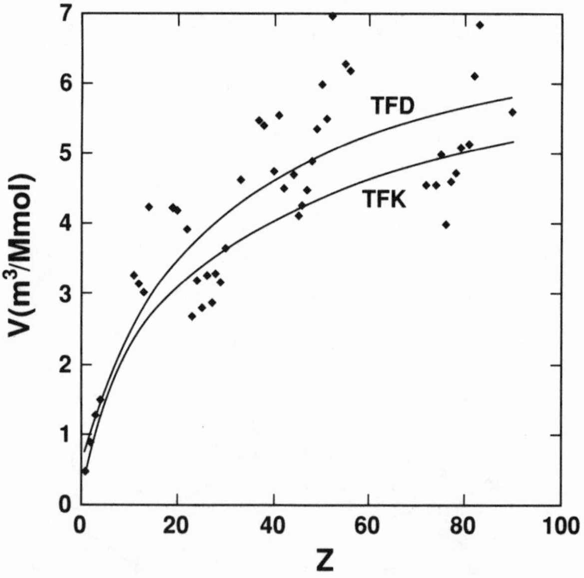

Fig. 17.2 Comparison of the estimated atomic volumes of the elements at 1 TPa with the predictions of the Thomas-Fermi-Dirac and Thomas-Fermi-Kalitkin models.

to the TFD curve is a few hundred GPa, equal to the bulk modulus of the least compressible elements.

The rapid convergence of the elemental volumes to a smooth function of Z in a few hundred GPa occurs because the elements with the largest volumes are also the most compressible. This correlation is best seen for the metals. In Fig. 17.3, the electron-radius parameter r_s for the interstitial electron gas computed from band-structure theory is correlated with the experimental bulk modulus for a group of metals[2]. The points lie close to the theoretical curve for the homogeneous electron gas, showing that metals behave much like the electron gas, and that under pressure, the elements with large r_s values will move up the curve toward those with smaller r_s values.

Another measure of bonding in the elements is the cohesive energy. A plot of E_{coh} vs. Z at $P = 0$ shows periodic behavior with large oscillations. The value of E_{coh} varies over a range of 10^4, which is much larger than the volume range. This function is less regular than other variables because it is the difference between the free-atom energy and the solid-state energy, both of which may fluctuate as a function of Z. However, the cohesive energy is not completely independent of V_o and B_o. The relationship between these variables is sketched in Fig. 17.4. Here it is clear that a large curvature in the

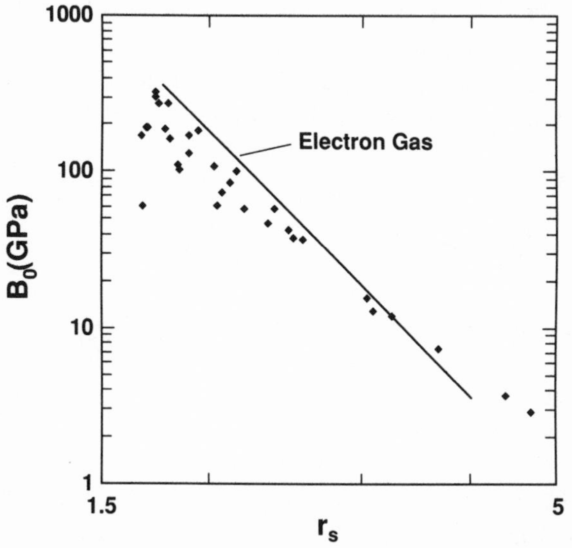

Fig. 17.3 A plot of the experimental bulk modulus of selected metals against the interstitial electron radius r_S computed from band-structure theory. The homogeneous electron-gas theory is superimposed on the plot.

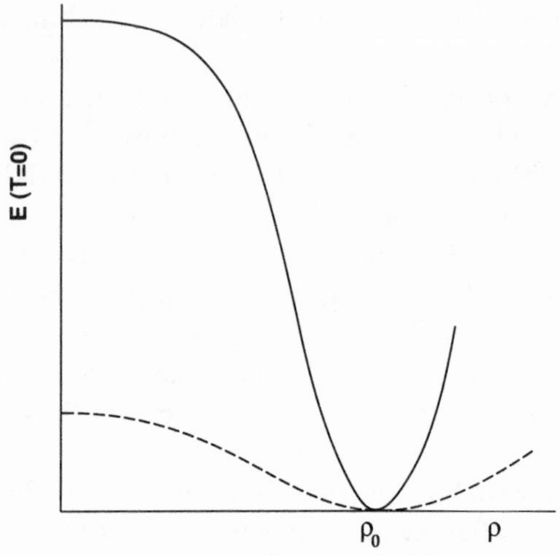

Fig. 17.4 A sketch showing the relation between the curvature of the $T = 0$ K energy isotherm at $\rho = \rho_o$ and the cohesive energy, $E(\rho = 0)$.

$T = 0$ energy isotherm, which corresponds to a large value of B_o, also corresponds to a large value of the $\rho = 0$ energy value, or E_{coh}. These parameters may be related formally in the McQueen-Marsh equation, $E_{coh} = 0.5\, A \times (C_o/S)^2 = 0.5\, A\, B_o V_o / S^2$, where A is the atomic weight, C_o is the bulk sound speed, and S is the slope of the u_s-u_p curve[3]. The two sides of this equation are plotted in Fig. 17.5, and the correlation is moderately good. Elements such as carbon show large deviations from the rule because vaporization leads to polyatomic species, not monatomic, as required by the definition of E_{coh}.

17.3 Crystal Structures

A more difficult property to explain is the most stable crystal structure of a solid element. The most stable structure of an element will depend on subtle details of the interatomic potential because that structure is competing with other structures of nearly equal free energy. Nevertheless, it is remarkable how many of the elements have the simple dense-packed structures fcc, hcp, or bcc. This stability reflects the geometric packing imposed by a smooth repulsive interatomic potential which favors fcc or hcp, or alternatively, the Madelung energy which favors bcc.

In Fig. 17.6 I show the crystal structures of the elements at $P = 0$. The solid rare gases and most of the metallic elements are dense-packed, while the

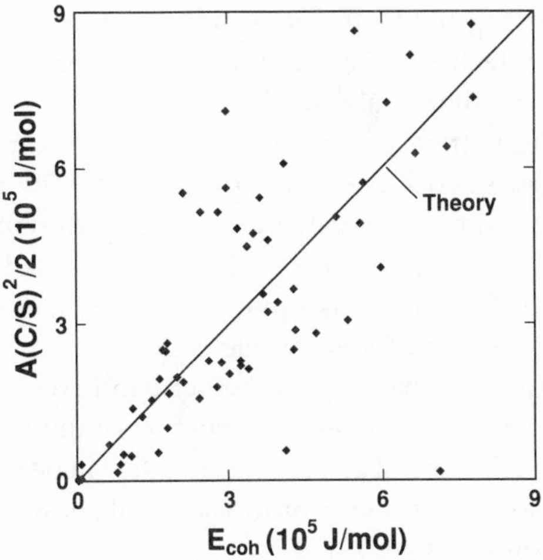

$$\mathbf{A(C/S)^2/2 \ (10^5 \ J/mol)}$$

Theory

$$\mathbf{E_{coh} \ (10^5 \ J/mol)}$$

Fig. 17.5 A test of the McQueen-Marsh relationship. The points are experimental data for the elements, and the straight line is the theoretical model.

P = 0 GPa

H																	He
X																	X
Li	Be											B	C	N	O	F	Ne
X	X											O	O	O	O	O	X
Na	Mg											Al	Si	P	S	Cl	Ar
X	X											X	O	O	O	O	X
K	Ca	Sc	Ti	V	Cr	Mn	Fe	Co	Ni	Cu	Zn	Ga	Ge	As	Se	Br	Kr
X	X	X	X	X	X	O	X	X	X	X	X	O	O	O	O	O	X
Rb	Sr	Y	Zr	Nb	Mo	Tc	Ru	Rh	Pd	Ag	Cd	In	Sn	Sb	Te	I	Xe
X	X	X	X	X	X	X	X	X	X	X	X	O	O	O	O	O	X
Cs	Ba	La	Hf	Ta	W	Re	Os	Ir	Pt	Au	Hg	Tl	Pb	Bi	Po	At	Rn
X	X	X	X	X	X	X	X	X	X	X	O	X	X	O	O	–	–
Fr	Ra	Ac															
–	X	X															

Ce	Pr	Nd	Pm	Sm	Eu	Gd	Tb	Dy	Ho	Er	Tm	Yb	Lu
X	X	X	X	X	X	X	X	X	X	X	X	X	X
Th	Pa	U	Np	Pu	Am	Cm	Bk	Cf	Es	Fm	Md	No	Lr
X	O	O	O	O	X	X	X	X	X	–	–	–	–

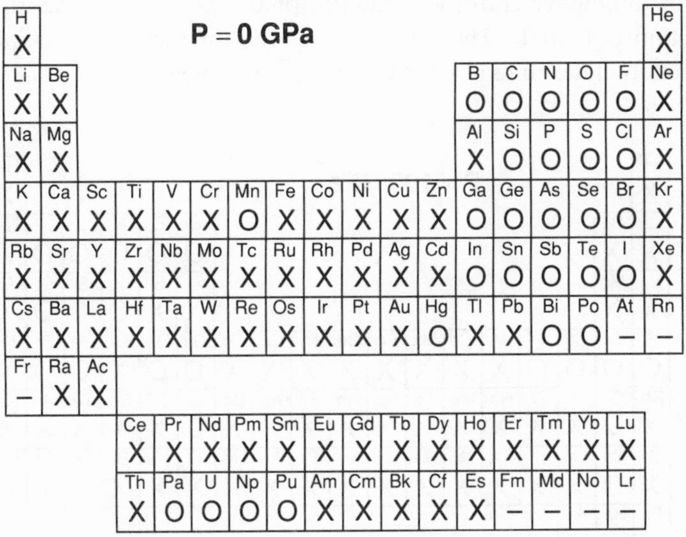

Fig. 17.6 The crystal structures of the elements at $P = 0$ GPa.
"X" = dense-packed structure and "O" = open-packed structure.

nonmetallic elements of the upper right corner of the periodic table are covalent and have more open structures. As pressure increases, covalent bonds are destabilized and the elements of the upper right corner evolve toward dense-packed structures. At the same time, the metals of the left side of the table, including the lanthanides and actinides, show an *s-d* or *s-f* electron transfer which stabilizes non-dense-packed structures such as tetragonal Cs IV or orthorhombic α-U.

At 100 GPa, the pattern of structures from measurement, theoretical calculation, and guesswork is shown in Fig. 17.7. The two "waves" of structure change are simultaneously moving across the table. Carbon may be the last covalent element to transform to a dense-packed metal, probably somewhere near 10 TPa. There is some evidence that at pressures above 100 GPa, where the *s-d* transfer is complete, the more open metallic structures will revert to dense packing. Whether band crossings at ultrahigh pressures will once again stabilize open structures is unknown, but we may speculate that at 1 TPa, almost all of the elements will be in metallic dense-packed lattices, and that further compression will allow phase transitions with only small volume changes.

Temperature also influences crystal structure, and a very common feature of many elemental phase diagrams is the appearance of the bcc phase before melting. As discussed in chapter 3, this may be explained in terms of the higher entropy of the bcc phase, given a sufficiently "soft" interaction potential. The "softness" of the interaction potential is approximately determined by the Grüneisen parameter γ.

P = 100 GPa

H X																	He X
Li X	Be X											B O	C O	N O	O O	F O	Ne X
Na X	Mg X											Al X	Si X	P O	S O	Cl O	Ar X
K O	Ca O	Sc O	Ti O	V X	Cr X	Mn X	Fe X	Co X	Ni X	Cu X	Zn X	Ga O	Ge O	As X	Se O	Br O	Kr X
Rb X	Sr O	Y X	Zr X	Nb X	Mo X	Tc X	Ru X	Rh X	Pd X	Ag X	Cd X	In O	Sn X	Sb X	Te X	I X	Xe X
Cs X	Ba O	La X	Hf X	Ta X	W X	Re X	Os X	Ir X	Pt X	Au X	Hg X	Tl X	Pb X	Bi X	Po –	At –	Rn –
Fr –	Ra –	Ac –															
			Ce O	Pr O	Nd O	Pm –	Sm O	Eu O	Gd X	Tb X	Dy X	Ho X	Er X	Tm X	Yb X	Lu X	
			Th X	Pa O	U O	Np O	Pu O	Am O	Cm O	Bk O	Cf O	Es –	Fm –	Md –	No –	Lr –	

Fig. 17.7 The crystal structures of the elements at $P = 100$ GPa. Many of these are theoretically predicted or are guesses. "X" = dense-packed structure and "O" = open-packed structure.

For a simple inverse-power potential, $\phi(r) = \varepsilon(\sigma/r)^n$, the bcc phase first appears at $n = 7.6$ or $\gamma = 1.6$. For smaller values of n or γ, the bcc phase will occupy more of the phase diagram. The experimental correlation of γ with the fraction of the phase diagram occupied by bcc is poor, however, because bcc may be stabilized, as in the transition metals, by static interatomic interactions unrelated to the lattice entropy. Nevertheless, the frequent appearance of a transition to bcc near melting implies that entropy is the dominant effect.

17.4 Melting and Vaporization

As with crystal structure, melting properties show systematic behavior across the periodic table. A plot of melting temperature vs. Z at $P = 0$ shows once again a periodic structure of alternating high and low values. The meaning of this periodicity is more clearly seen if RT_m is plotted against E_{coh}, as in Fig. 17.8. There is a strong correlation between these two variables. The theoretical explanation for this relation may be derived from the melting of repulsive potential particles in an attractive mean field, illustrated in Fig. 17.9. As the depth of the mean field (i.e., the cohesive energy) is

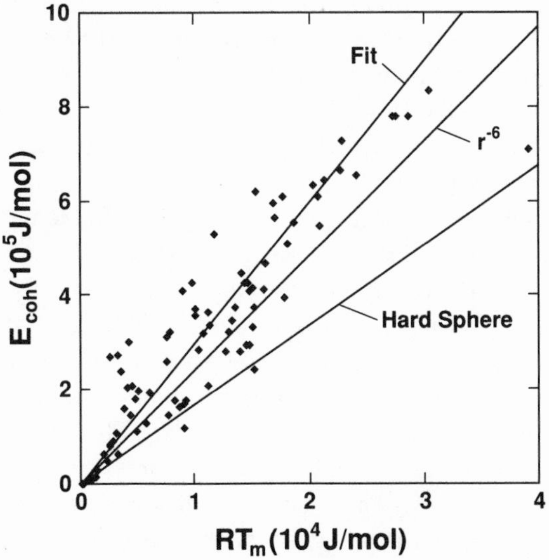

Fig. 17.8 Comparison of the experimental melting "energy" RT_m with the cohesive energy for the elements. Predictions of two theories based on repulsive cores in a mean field are shown.

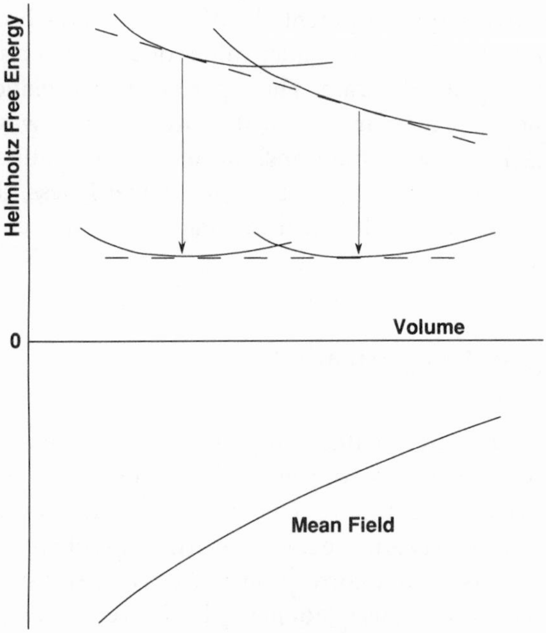

Fig. 17.9 A sketch of the free energies of hard-core solid and liquid and the effect of an attractive mean field.

increased, the melting temperature at $P = 0$ increases according to a linear relation. This relation depends on the steepness of the repulsive interaction and the form of the attractive mean field. The experimental correlation in Fig. 17.8 is best fitted with soft-core potentials and mean-field attractions varying more rapidly than the first power of the density.

As pressure increases, the melting temperature normally increases also. In those cases where the melting slope dP/dT is found to be negative, there will be a triple point at higher pressure and a reversal of sign of dP/dT. A plot of T_m vs. Z for $P = 10$ GPa still shows strong periodicity, but the lowest melting temperatures have increased rapidly, so that the ratio of highest to lowest is much smaller than at $P = 0$. Shock-wave experiments and theoretical calculations have been providing new data on melting in the 100 GPa pressure region, and it is now possible to consider the evolution of the melting curves of the elements to very high pressures. This is shown in Fig. 17.10 for 9 elements. These curves indicate a remarkable increase in melting temperature of the rare gases with pressure. The melting curve of Ar crosses those of Fe, Al, and Xe by 100 GPa. It appears that at pressures above 100 GPa, the melting curves of elements like Ne and Ar may cross many other melting curves until they themselves metallize.

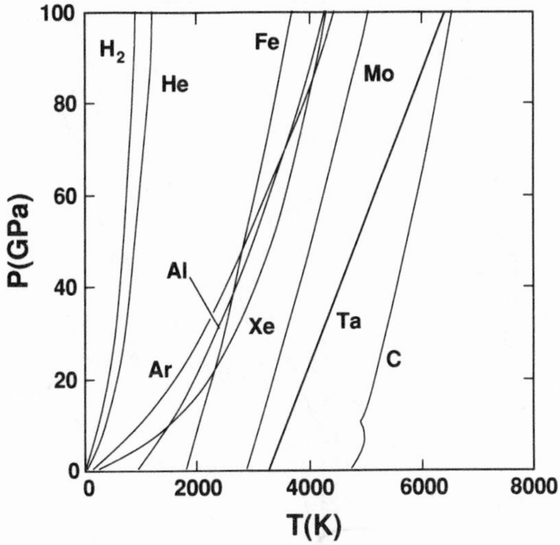

Fig. 17.10 The theoretically estimated melting curves of 9 elements to 100 GPa.

At extreme pressures, atoms are completely pressure-ionized and their melting curves will approach those of the one-component plasma with nuclear charge Z. These melting curves will then form a monotonic series with temperatures increasing with Z, as shown in Fig. 17.11. The lower-pressure melting curves are expected to cross and recross one another as this limit is approached. In contrast with the behavior of the atomic volumes, the monotonicity of melting becomes visible only at pressures and temperatures far beyond the experimental range.

At low pressures and at temperatures above the melting point we find the liquid-vapor region and the critical point. The critical points of about 20 elements have been measured (Table 16.2), and it is useful to examine these data as a group. As with melting, there is a natural correlation between T_c and E_{coh}. A theoretical correlation can be taken from the hard-sphere theory of the critical point, in which the two parameters are related according to Eq. (16.9b). This correlation is plotted in Fig. 17.12 and is found to be a good fit to the collected critical point data.

When insulating solids are heated at low pressure, they enter the vapor phase, which is also insulating. At sufficiently high temperature, ionization of the vapor will create a conducting plasma. Similarly, under pressure the electrons in the valence band will eventually delocalize and create a conducting metallic state. This means that the insulator-to-conductor transition will form a closed loop in the P-T phase diagram. For an element which is metallic at $P = 0$, the nonmetallic state is found only in the vapor.

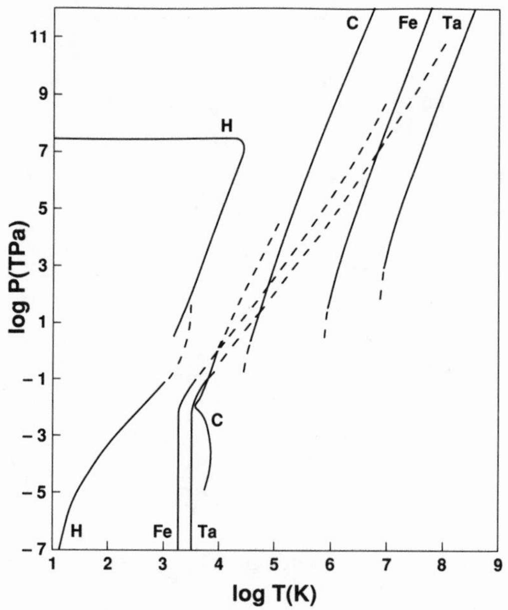

Fig. 17.11 The theoretically determined ultrahigh-pressure melting curves of 4 elements.

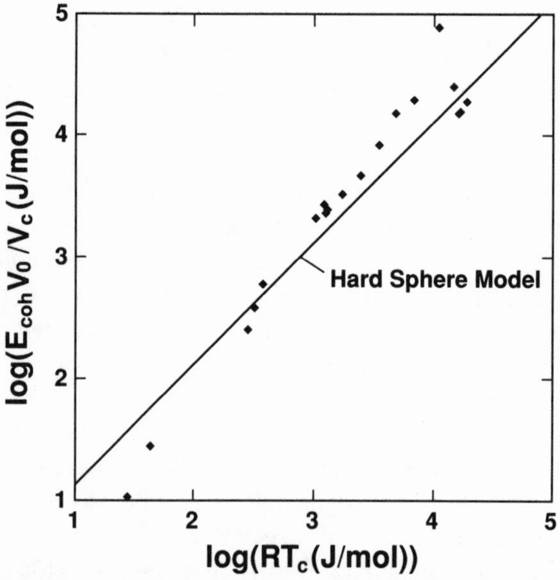

Fig. 17.12 A comparison of experimental critical point data with the cohesive energy. The hard-sphere theory is given by the straight line.

Here the metal-nonmetal transition occurs in the solid only at artificial negative pressures. Some schematic phase diagrams illustrating these trends are shown in Fig. 17.13. The implication of this generality is a kind of symmetry between the temperature and pressure axes.

17.5 Corresponding States

The principle of corresponding states, which relates the equation of state of one substance to another when both share a common two-parameter potential function, is expected to hold approximately for elements belonging to the same column in the periodic table. This has been shown to be true, for example, for rare gases at low pressure and for alkali metals at the critical point. However, the very high pressures now attainable in the DAC allow us to examine the principle more critically. Here I review some of the common patterns found among the elements.

The elements of Groups I and II may be divided into two blocks: the 1*s* and 2*s* elements (Li, Be, Na, Mg) which have no nearby *d* band; and the others, which do. The heavier members of the groups are highly compressible and have many phase transitions, which is the result of *s-d* electron transfer. Band-structure calculations describe this phenomenon in accurate detail for Cs and relate it to the phase diagram.

The heavier Group I elements clearly show a tendency for the same transition to occur at lower pressure in the heavier element. This results

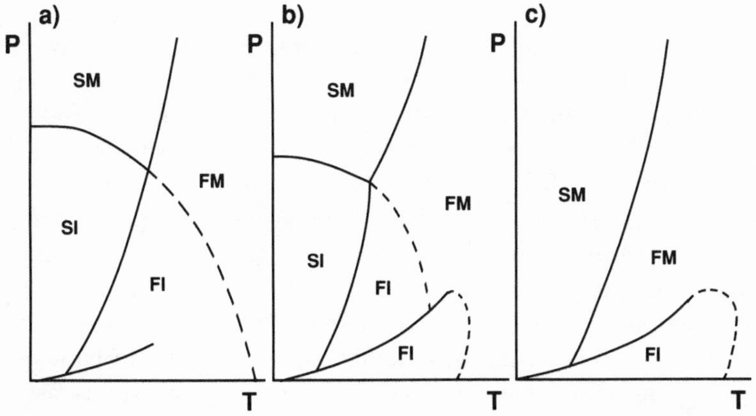

Fig. 17.13 Schematic phase diagrams showing insulator-metal transitions in three elements, a) a low-boiling insulator, b) a covalent insulator, and c) a metal. SI = solid insulator, SM = solid metal, FI = fluid insulator, FM = fluid metal.

from the larger size of the heavier atom which leads to lower energies, smaller band gaps, and a lower transition pressure. The Group II's have a slightly different pattern in which the phase diagrams appear to be shifted with respect to one another rather than scaled or compressed. Melting and solid-solid phase transitions at $T > 0$ in these elements are well represented by statistical-mechanical theory using pseudopotentials.

There is no quantitatively accurate scaling rule for the Group I and II *P-T* phase diagrams, but the phase transitions governed by the *s-d* transfer have approximately constant values of the radius ratio R_A/R_I, where R_A is the atomic radius and R_I is the ionic radius.

The Group III elements show some unusual structures which result from the undulations of the pair potential. These structures can be related rather simply to the pseudopotential parameters, as shown in Fig. 7.9.

The Group IV, V, and VI elements show covalent bonding and open metallic structures which imply strong electron correlation in the valence band. The open simple-cubic-like structures are the result of the *p* orbitals which favor bond angles of 90°. These groups also show a tendency toward lower transition pressures for the heavier members of the group as in Group I.

Also very apparent in these groups are metallization and simplification of the crystal structures in the metallic state under compression. Band-structure theory has been very successful in predicting metallization in such elements as Si, Ge, and P, as well as the observed sequence of crystal structures. This gives us confidence in the band-structure predictions for elements like C, where the extreme transition pressures make experiments currently impossible.

The Group VII elements are all diatomic and show closely packed structures. Under pressure, it is expected that these diatomic solids will dissociate, as in Br_2 and I_2, to low-symmetry monatomic metals, and that the symmetry will increase toward close-packing under further compression.

For the diatomic elements H_2, N_2, O_2, F_2, Cl_2, Br_2, and I_2, there is now a qualitative understanding, based on empirical intermolecular potentials, of the observed zero-pressure solid phases. The important variables appear to be the molecular electrostatic multipoles and the molecular shape. A general diatomic phase diagram is shown in Fig. 11.6.

The metallization process in diatomics may occur either by band overlap or by dissociation and structural change. The only unambiguous case is that of I_2, which metallizes by band overlap while still diatomic. Preliminary data on O_2 and Br_2 suggest that band overlap is also the mechanism for metallization in these elements. High-pressure experiments on H_2 and N_2 have not yet convincingly metallized these elements.

The Group VIII (rare gas) elements have been intensively studied because of their closed-shell electronic structures and the resulting weak interaction potentials. The Group VIII phase diagrams at low temperatures are thus well understood in terms of simple two-body interaction potentials. Since the interaction potentials in the neighborhood of the attractive minimum all have nearly the same analytic form, the corresponding-states principle is accurately obeyed and all of the phase diagrams (except He) can be reduced to a single universal diagram by scaling P and T at low pressure.

At higher pressures, the interatomic forces become more complex as many-body contributions increase in strength. The solids also go through solid-solid phase transitions and metallization, which are governed by the details of the band structures. This high-pressure behavior does not obey a simple scaling relation, so corresponding states must break down. This can be seen in the predicted crossing of the Kr and Xe melting curves in Fig. 12.11.

The transition metals show the pattern hcp-bcc-hcp-fcc in moving from left to right across the rows. This progression is accurately predicted by band-structure theory. The structures are determined by the d occupation number. The principal anomaly lies in the effects of 3d magnetism. Generalized Pseudopotential Theory shows that for the transition metals the effective potential has important three- and higher-body interactions as well as the two-body contribution. This complicates the description of the condensed phases, but calculations of phonon spectra and melting curves nonetheless show good agreement with experiments.

Because of the very high bulk moduli of the transition metals, pressure-induced phase transitions are expected only at hundreds of GPa pressures. These transitions are beginning to be explored by DAC and shock-wave techniques. The general tendency, according to band-structure calculations, is that pressure will fill the d-band and hence drive the structures of a given column toward the next column to the right. For the later elements in the series, the tendency is reversed, and the d-band empties slightly under pressure.

The lanthanides lie between Ba ($6s^2$) and Hf ($6s^2 5d^2 4f^{14}$) and they represent the filling of the 4f shell. The 4f's fill rather small orbitals and, except for Ce, do not participate in metallic bonding at low pressures. Thus the lanthanides look like transition metals with between 1 and 2 d electrons. Band-structure theory shows that the sequence of phase transitions observed in the lanthanides results from an s-d electron transfer. The d-occupation number increases with compression or with decreasing Z, owing to ion-core size effects. The progression of *d*-electron number in the trivalent lanthanides is so smooth that the phase diagrams can be fitted into a single general diagram, shown in Fig. 14.21. Very strong compression of the lanthanides forces the 4*f*'s into the valence band, which stabilizes open structures like orthorhombic α-U.

The actinides are analogous to the lanthanides, except that the 5*f* electrons have larger orbitals and they more actively participate in the metallic bonding. Hence Pa, U, Np, and Pu all show non-close-packed crystal structures because of 5*f* bonding. With Am, the lanthanide-type behavior begins. There is a rapid shrinkage and energy lowering of the 5*f* orbitals with increasing Z, so that the heaviest actinides have the configuration $5f^n7s^2$ and are divalent rather than trivalent.

The lanthanide-actinide relationship is more complex than in the related columns of *sp* or *d* elements. The trivalent lanthanides may be correlated with the heavy actinides at $P = 0$ as in Fig. 15.8, or the lanthanides at ca. 100 GPa may be correlated with the actinides at $P = 0$.

The advance of experimental techniques at very high pressures has made it harder to make simplistic statements about generalized phase diagrams for groups of the elements. It is evident that the periodic behavior of the elements and their correspondences are "low-pressure" phenomena, and that at pressures beyond the present experimental range, the behavior of the elements will begin to have a monotonic Z dependence. Corresponding-states rules must therefore apply to strictly limited pressure ranges, and must eventually give way to scaling laws reflecting the behavior of pressure-ionized matter.

17.6 Prospects

We can expect further increases in attainable pressures and temperatures in DAC experiments. The upper limit of pressure for the DAC is likely to be the pressure of the first phase transition from diamond to another phase, which is theoretically predicted at 1.1 TPa. At present, laser heating can achieve ca. 5000 K at 100 GPa, and it is possible that more powerful lasers could yield still higher temperatures. Higher pressures in shock-wave experiments are less interesting, because the rapidly rising temperature of the shocked material causes melting, and all Hugoniot states above this point are in the liquid phase. However, the optical-analyzer technique is yielding much new data on phase transitions in the very incompressible transition metals, and new measurements of shock temperatures in metals promise to yield phase-diagram data in the hundreds of GPa region.

Theoretical understanding of phase behavior is also making rapid progress. There is good confidence in the predictions of total-energy band-structure calculations for various crystal structures of the elements. There are still some elements like Fe and Ce where strong electron-correlation effects make the local-density approximation inadequate, but work on nonlocal theories has already begun to rectify this. For nonzero temperatures, total-energy calculations may be used to obtain selected phonon frequencies by distorted-cell calculations, but this is an extremely laborious effort. More promising is the Generalized Pseudopotential Theory, which yields two-body and higher n-body interaction potentials for metals. Calculation of phonon frequencies allows prediction of the stability of high-temperature solid phases and the *P-T* phase boundaries between them. Liquid and solid theories are now sufficiently accurate that, if supplied with good interaction potentials, they can produce an accurate melting curve.

The MC and MD computer-simulation methods are rapidly increasing in importance, and are beginning to constitute a "third branch" of condensed-matter physics, distinct from experiment and standard quantum- and statistical-mechanical theory. The Car-Parrinello scheme, which combines *ab initio* treatment of electrons with classical nuclear motion, has been very informative about chemical bonding in covalent solids and liquids at high temperature.

For diatomic solids and liquids, classical MD calculations have revealed new data about correlated molecular motions. The use of variable-boundary-condition methods has permitted "natural" searches for the most stable crystal structures, and new free-energy calculations have permitted accurate prediction of phase transitions. These simulations reveal information about atomic configurations and motions currently unobtainable from experiment and standard theory, and they thus contribute importantly to our understanding of condensed-matter physics.

As we obtain more data on the solid and liquid phases of the elements, we discover more correlations among the measured properties which bring out new relationships among the elements. However, we also discover exceptions to the rules which emphasize the uniqueness of each element. The rich structure of the periodic table is revealed in the balancing of these two tendencies.

References

1. L. V. Al´tshuler and A. A. Bakanova, Usp. Fiz. Nauk **96**, 193 (1968) [Sov. Phys. Usp. **11**, 678 (1969)].
2. V. L. Moruzzi, J. F. Janak, and A. R. Williams, *Calculated Electronic Properties of Metals* (Pergamon, New York, 1978).
3. R. G. McQueen and S. P. Marsh, J. Appl. Phys. **33**, 654 (1962).

APPENDIXES

The following tables list the atomic weight, molar volume, bulk modulus, cohesive energy, and melting temperature for the elements. I have not attempted to obtain the best numbers from the primary literature, but have used published compilations where possible. The reference for each element is given in the right-hand column.

APPENDIX A
Atomic Weight

These numbers are from recent compilations and should be very accurate. For the artificial elements, I have chosen the atomic weight of the longest-lived isotope.

Z	Element	Atomic Weight	Reference
1	H	1.008	1,2
1	D	2.014	1,2
2	He-3	3.016	1
2	He-4	4.003	1
3	Li	6.941	1
4	Be	9.012	1
5	B	10.811	1
6	C	12.011	1
7	N	14.007	1
8	O	15.999	1
9	F	18.998	1
10	Ne	20.180	1
11	Na	22.990	1
12	Mg	24.305	1
13	Al	26.982	1
14	Si	28.086	1
15	P	30.974	1
16	S	32.066	1
17	Cl	35.453	1
18	Ar	39.948	1
19	K	39.098	1
20	Ca	40.078	1
21	Sc	44.956	1

(continued)

Appendix A (continued)

Z	Element	Atomic Weight	Reference
22	Ti	47.88	1
23	V	50.942	1
24	Cr	51.996	1
25	Mn	54.938	1
26	Fe	55.847	1
27	Co	58.933	1
28	Ni	58.69	1
29	Cu	63.546	1
30	Zn	65.39	1
31	Ga	69.723	1
32	Ge	72.61	1
33	As	74.922	1
34	Se	78.96	1
35	Br	79.904	1
36	Kr	83.80	1
37	Rb	85.468	1
38	Sr	87.62	1
39	Y	88.906	1
40	Zr	91.224	1
41	Nb	92.906	1
42	Mo	95.94	1
43	Tc	97.907	2
44	Ru	101.07	1
45	Rh	102.906	1
46	Pd	106.42	1
47	Ag	107.868	1
48	Cd	112.411	1
49	In	114.82	1
50	Sn	118.710	1
51	Sb	121.75	1
52	Te	127.60	1
53	I	126.904	1
54	Xe	131.29	1
55	Cs	132.905	1

(continued)

Appendix A (continued)

Z	Element	Atomic Weight	Reference
56	Ba	137.327	1
57	La	138.906	1
58	Ce	140.115	1
59	Pr	140.908	1
60	Nd	144.24	1
61	Pm	144.913	2
62	Sm	150.36	1
63	Eu	151.965	1
64	Gd	157.25	1
65	Tb	158.925	1
66	Dy	162.50	1
67	Ho	164.930	1
68	Er	167.26	1
69	Tm	168.934	1
70	Yb	173.04	1
71	Lu	174.967	1
72	Hf	178.49	1
73	Ta	180.948	1
74	W	183.85	1
75	Re	186.207	1
76	Os	190.2	1
77	Ir	192.22	1
78	Pt	195.08	1
79	Au	196.966	1
80	Hg	200.59	1
81	Tl	204.383	1
82	Pb	207.2	1
83	Bi	208.980	1
84	Po	208.982	2
85	At	209.987	2
86	Rn	222.018	2
87	Fr	223.020	2
88	Ra	226.025	2

(continued)

Appendix A (continued)

Z	Element	Atomic Weight	Reference
89	Ac	227.028	2
90	Th	232.038	1
91	Pa	231.036	2
92	U	238.029	1
93	Np	237.048	2
94	Pu	244.064	2
95	Am	243.061	2
96	Cm	247.070	2
97	Bk	247.070	2
98	Cf	251.080	2
99	Es	252.083	2
100	Fm	257.095	2

References

1. J. R. DeLaeter, J. Phys. Chem. Ref. Data **17**, 1791 (1988).
2. R. J. Howerton, Lawrence Livermore National Laboratory Report UCID-20487, 1985.

APPENDIX B
Molar Volume

For low-melting-temperature elements, the volumes are taken from measurements at the lowest possible temperature, which is shown in parentheses. For higher-melting elements, the volumes refer to room temperature. For the diatomic elements H_2, D_2, N_2, O_2, F_2, Cl_2, Br_2, and I_2, the volume is the diatomic molar volume; for all other elements the atomic volume is used.

Z	Element	Volume (m^3/Mmol)	Reference
1	H	23.06(0 K)	1
1	D	19.93(0 K)	1
2	He-3	36.84(0 K)	2
2	He-4	27.58(0 K)	2
3	Li	12.68(4 K)	3
4	Be	4.85	4
5	B	4.39	4
6	C	5.29	4
7	N	27.14(8 K)	5
8	O	20.81(7 K)	6
9	F	19.29(23 K)	7
10	Ne	13.39(0 K)	8
11	Na	22.66(4 K)	9
12	Mg	14.00	4
13	Al	10.00	4
14	Si	12.06	4
15	P	11.44	4
16	S	15.53	4

(Continued)

Appendix B (continued)

Z	Element	Volume (m^3/Mmol)	Reference
17	Cl	33.16(22 K)	10
18	Ar	22.56(0 K)	8
19	K	43.20(0 K)	9
20	Ca	25.94(0 K)	11
21	Sc	15.00	4
22	Ti	10.64	4
23	V	8.32	4
24	Cr	7.23	4
25	Mn	7.35	4
26	Fe	7.09	4
27	Co	6.67	4
28	Ni	6.59	4
29	Cu	7.11	4
30	Zn	9.16	4
31	Ga	11.65(4 K)	12
32	Ge	13.63	4
33	As	12.95	4
34	Se	16.42	4
35	Br	38.40(5 K)	10
36	Kr	27.10(0 K)	8
37	Rb	52.60(4 K)	9
38	Sr	33.31(0 K)	11
39	Y	19.88	4
40	Zr	14.02	4
41	Nb	10.83	4
42	Mo	9.38	4
43	Tc	8.63	4
44	Ru	8.17	4
45	Rh	8.28	4
46	Pd	8.56	4
47	Ag	10.27	4
48	Cd	13.00	4
49	In	15.76	4

(continued)

Appendix B (continued)

Z	Element	Volume (m^3/Mmol)	Reference
50	Sn	16.29	4
51	Sb	18.19	4
52	Te	20.46	4
53	I	49.26(110 K)	12
54	Xe	34.74(0 K)	8
55	Cs	66.42(4 K)	3
56	Ba	37.79(0 K)	11
57	La	22.60	13
58	Ce	20.70	13
59	Pr	20.80	13
60	Nd	20.58	13
61	Pm	20.24	13
62	Sm	20.00	13
63	Eu	28.98	13
64	Gd	19.90	13
65	Tb	19.31	13
66	Dy	19.00	13
67	Ho	18.75	13
68	Er	18.45	13
69	Tm	18.12	13
70	Yb	24.84	13
71	Lu	17.78	13
72	Hf	13.44	4
73	Ta	10.85	4
74	W	9.47	4
75	Re	8.86	4
76	Os	8.42	4
77	Ir	8.52	4
78	Pt	9.09	4
79	Au	10.21	4
80	Hg	13.58(77 K)	12
81	Tl	17.22	4
82	Pb	18.26	4

(continued)

Appendix B (continued)

Z	Element	Volume (m^3/Mmol)	Reference
83	Bi	21.31	4
84	Po	22.97	4
85	At	—	
86	Rn	50.50	4
87	Fr	—	
88	Ra	41.09	4
89	Ac	22.55	4
90	Th	19.80	4
91	Pa	14.98	14
92	U	12.49	4
93	Np	11.59	4
94	Pu	12.04	12
95	Am	17.63	4
96	Cm	18.05	4
97	Bk	16.84	4
98	Cf	16.50	4
99	Es	28.52	4
100	Fm	—	

References

1. P. C. Souers, *Hydrogen Properties for Fusion Energy* (University of California Press, Berkeley, Los Angeles, London, 1986) p. 78.
2. R. K. Crawford, in *Rare Gas Solids* M. L. Klein and J. A. Venables, eds. (Academic, London, 1977) vol. 2, chap. 11, pp. 674–675.
3. M. S. Anderson and C. A. Swenson, Phys. Rev. B **31**, 668 (1985).
4. C. N. Singman, J. Chem. Educ. **61**, 137 (1984).
5. T. A. Scott, Phys. Repts. **27**, 89 (1976).
6. I. N. Krupskii, A. I. Prokhvatilov, Yu. A. Freiman, and A. I. Erenburg, Fiz. Nizk. Temp. **5**, 271 (1979) [Sov. J. Low Temp. Phys. **5**, 130 (1979)].
7. L. Meyer, C. S. Barrett, and S. C. Greer, J. Chem. Phys. **49**, 1902 (1968).
8. M. S. Anderson and C. A. Swenson, J. Phys. Chem. Solids **36**, 145 (1975).
9. M. S. Anderson and C. A. Swenson, Phys. Rev. B **28**, 5395 (1983).
10. B. M. Powell, K. M. Heal, and B. H. Torrie, Mol. Phys. **53**, 929 (1984).
11. M. S. Anderson, C. A. Swenson, and D. T. Peterson, Phys. Rev. B **41**, 3329 (1990).
12. J. Donohue, *The Structures of the Elements* (Wiley, New York, 1974).

13. B. J. Beaudry and K. A. Gschneidner, Jr., in *Handbook on the Physics and Chemistry of the Rare Earths* K. A. Gschneidner, Jr. and L. Eyring, eds. (North-Holland, Amsterdam, 1978) vol. 1, chap. 2.
14. J. W. Ward, P. D. Kleinschmidt, and D. E. Peterson, in *Handbook on the Physics and Chemistry of the Actinides* A. J. Freeman and C. Keller, eds. (North-Holland, Amsterdam, 1986) vol. 4, chap. 7.

APPENDIX C
Bulk Modulus

For low-melting-temperature elements, the bulk moduli are taken from measurements at the lowest possible temperature, which is shown in parentheses. For higher-melting elements, the room temperature value is listed.

Z	Element	Bulk Modulus (GPa)	Reference
1	H	0.186(0 K)	1
1	D	0.335(0 K)	2
2	He-3	0.0027(0 K)	3
2	He-4	0.0078(0 K)	4
3	Li	12.6(0 K)	5
4	Be	110	6,7
5	B	196	7,8
6	C	35.8	9
7	N	2.16(8 K)	10
8	O	3.60(7 K)	11
9	F	—	
10	Ne	1.10(0 K)	12
11	Na	7.34(4 K)	13
12	Mg	34.1	6,7
13	Al	72.8	6,7
14	Si	97.7	6,7
15	P	32.5	14
16	S	7.61	15
17	Cl	14.2	16

(continued)

Appendix C (continued)

Z	Element	Bulk Modulus (GPa)	Reference
18	Ar	2.86(0 K)	12
19	K	3.70(4 K)	13
20	Ca	18.4(0 K)	17
21	Sc	54.6	6,7
22	Ti	106	6,7
23	V	155	6,7
24	Cr	160	6,7
25	Mn	90.4	6,7
26	Fe	163	6,7
27	Co	186	6,7
28	Ni	179	6,7
29	Cu	133	6,7
30	Zn	64.8	6,7
31	Ga	56.9	6,7
32	Ge	74.9	6,7
33	As	63.1	6,7
34	Se	9.1	7
35	Br	12.2	16
36	Kr	3.34(0 K)	12
37	Rb	2.92(4 K)	13
38	Sr	12.4(0 K)	17
39	Y	41.0	6,7
40	Zr	94.9	6,7
41	Nb	169	6,7
42	Mo	261	6,7
43	Tc	—	
44	Ru	303	6,7
45	Rh	282	6,7
46	Pd	189	6,7
47	Ag	98.8	6,7
48	Cd	49.8	6,7
49	In	39.3	6,7
50	Sn	55.0	6,7

(continued)

Appendix C (continued)

Z	Element	Bulk Modulus (GPa)	Reference
51	Sb	41.1	6,7
52	Te	23.3	6,7
53	I	13.6	16
54	Xe	3.65(0 K)	12
55	Cs	2.10(4 K)	5
56	Ba	9.30(0 K)	17
57	La	26.6	6,7
58	Ce	20.8	6,7
59	Pr	28.7	6,7
60	Nd	32.7	6,7
61	Pm	38.0	18
62	Sm	36.0	6,7
63	Eu	16.2	6,7
64	Gd	37.6	6,7
65	Tb	38.4	6,7
66	Dy	40.5	6,7
67	Ho	40.4	6,7
68	Er	45.0	6,7
69	Tm	45.6	6,7
70	Yb	14.6	6,7
71	Lu	47.4	6,7
72	Hf	108	6,7
73	Ta	191	6,7
74	W	308	6,7
75	Re	360	6,7
76	Os	—	
77	Ir	358	6,7
78	Pt	277	6,7
79	Au	166	6,7
80	Hg	28.2	7
81	Tl	33.7	6,7
82	Pb	41.7	6,7

(continued)

Appendix C (continued)

Z	Element	Bulk Modulus (GPa)	Reference
83	Bi	33.2	6,7
84	Po	—	
85	At	—	
86	Rn	—	
87	Fr	—	
88	Ra	—	
89	Ac	—	
90	Th	56.9	6,7
91	Pa	157	19
92	U	111	6,7
93	Np	118	20
94	Pu	47.6	21
95	Am	45.0	22
96	Cm	25.0	23
97	Bk	30.0	24
98	Cf	50.0	25
99	Es	—	
100	Fm	—	

References

1. S. N. Ishmaev, I. P. Sadikov, A. A. Chernyshov, B. A. Vindryaevskii, V. A. Sukhoparov, A. S. Telepnev, and G. V. Kobelev, Zh. Eksp. Teor. Fiz. **84**, 394 (1983) [Sov. Phys. JETP **57**, 228 (1983)].
2. S. N. Ishmaev, I. P. Sadikov, A. A. Chernyshov, B. A. Vindryaevskii, V. A. Sukhoparov, A. S. Telepnev, G. V. Kobelev, and R. A. Sadykov, Zh. Eksp. Teor. Fiz. **89**, 1249 (1985) [Sov. Phys. JETP **62**, 721 (1985)].
3. J. Wilks, *The Properties of Liquid and Solid Helium* (Oxford University Press, Oxford, 1967) p. 673.
4. J. Wilks, Ref. 3, p. 667.
5. M. S. Anderson and C. A. Swenson, Phys. Rev. B **31**, 668 (1985).
6. M. W. Guinan and D. J. Steinberg, J. Phys. Chem. Solids **35**, 1501 (1974).
7. K. A. Gschneidner, Jr., Solid State Phys. **16**, 275 (1964).
8. S. P. Marsh, ed., *LASL Shock Hugoniot Data* (University of California Press, Berkeley, Los Angeles, London, 1980).

9. Y. X. Zhao and I. L. Spain, Phys. Rev. B **40**, 993 (1989).
10. T. A. Scott, Phys. Repts. **27**, 89 (1976).
11. I. N. Krupskii, A. I. Prokhvatilov, Yu. A. Freiman, and A. I. Erenburg, Fiz. Nizk. Temp. **5**, 271 (1979) [Sov. J. Low Temp. Phys. **5**, 130 (1979)].
12. C. A. Swenson, in *Rare Gas Solids*, M. L. Klein and J. A. Venables, eds. (Academic, London, 1977) vol. 2, chap. 13.
13. M. S. Anderson and C. A. Swenson, Phys. Rev. B **28**, 5395 (1983).
14. L. Cartz, S. R. Srinavasa, R. J. Riedner, J. D. Jorgensen, and T. G. Worlton, J. Chem. Phys. **71**, 1718 (1979).
15. G. A. Saunders, Y. K. Yogurtçu, J. E. Macdonald, and G. S. Pawley, Proc. R. Soc. Lond. A **407**, 325 (1986).
16. E.-Fr. Düsing, W. A. Grosshans, and W. B. Holzapfel, J. Phys. (Paris) **45**, C8-203 (1984).
17. M. S. Anderson, C. A. Swenson, and D. T. Peterson, Phys. Rev. B **41**, 3329 (1990).
18. R. G. Haire, S. Heathman, and U. Benedict, preprint.
19. U. Benedict, J. C. Spirlet, C. Dufour, I. Birkel, W. B. Holzapfel, and J. R. Peterson, J. Magn. Magn. Mat. **29**, 287 (1982).
20. S. Dabos, C. Dufour, U. Benedict, and M. Pagés, J. Magn. Magn. Mat. **63&64**, 661 (1987).
21. M. D. Merz, J. H. Hammer, and H. E. Kjarmo, in *Plutonium and Other Actinides*, H. Blank and R. Lindner, eds. (North-Holland, Amsterdam, 1976) p. 567.
22. U. Benedict, J. P. Itié, C. Dufour, S. Dabos, and J. C. Spirlet, in *Americium and Curium Chemistry and Technology*, N. M. Edelstein, J. D. Navratil, and W. M. Schulz, eds. (D. Reidel, Boston, 1985) p. 213.
23. J. Akella, personal communication.
24. R. G. Haire, J. R. Peterson, U. Benedict, and C. Dufour, J. Less-Common Met. **102**, 119 (1984).
25. J. R. Peterson, U. Benedict, C. Dufour, I. Birkel, and R. G. Haire, J. Less-Common Met. **93**, 353 (1983).

APPENDIX D
Cohesive Energy

The cohesive energy is the heat of vaporization at 0 K. For the diatomic elements H_2, D_2, N_2, O_2, F_2, Cl_2, Br_2, and I_2, the cohesive energy refers to 1 mole of molecules. For all other elements, the cohesive energy refers to one mole of atoms in the gaseous state.

Z	Element	Cohesive Energy (kJ/mol)	Reference
1	H	0.747	1
1	D	1.104	1
2	He-3	0.0206	2
2	He-4	0.0596	2
3	Li	157.7	3
4	Be	320	4
5	B	566	4
6	C	711	4
7	N	6.92	5
8	O	8.66	6
9	F	9.18	6
10	Ne	1.887	2
11	Na	107.8	3
12	Mg	145	4
13	Al	327	4
14	Si	451	4
15	P	331	4
16	S	276	7
17	Cl	30.0	6

(continued)

Appendix D (continued)

Z	Element	Cohesive Energy (kJ/mol)	Reference
18	Ar	7.73	2
19	K	89.9	3
20	Ca	178	4
21	Sc	376	4
22	Ti	467	4
23	V	511	4
24	Cr	395	4
25	Mn	282	4
26	Fe	413	4
27	Co	427	4
28	Ni	428	4
29	Cu	336	4
30	Zn	130	4
31	Ga	271	4
32	Ge	372	4
33	As	302	4
34	Se	206	7
35	Br	46.3	6
36	Kr	11.1	2
37	Rb	82.2	3
38	Sr	164	7
39	Y	424	4
40	Zr	607	4
41	Nb	718	4
42	Mo	656	4
43	Tc	688	4
44	Ru	650	4
45	Rh	552	4
46	Pd	376	4
47	Ag	284	4
48	Cd	112	4
49	In	243	4
50	Sn	301	4

(continued)

Appendix D (continued)

Z	Element	Cohesive Energy (kJ/mol)	Reference
51	Sb	264	4
52	Te	196	4
53	I	65.5	4
54	Xe	15.84	2
55	Cs	78.0	3
56	Ba	183	4
57	La	431	4
58	Ce	423	4
59	Pr	357	4
60	Nd	328	4
61	Pm	—	
62	Sm	206	4
63	Eu	175	4
64	Gd	399	4
65	Tb	391	4
66	Dy	293	4
67	Ho	303	4
68	Er	318	4
69	Tm	233	4
70	Yb	153	4
71	Lu	428	4
72	Hf	619	4
73	Ta	781	4
74	W	848	4
75	Re	774	4
76	Os	788	4
77	Ir	668	4
78	Pt	564	4
79	Au	368	4
80	Hg	60.3	4
81	Tl	182	4
82	Pb	196	4
83	Bi	210	4

(continued)

Appendix D (continued)

Z	Element	Cohesive Energy (kJ/mol)	Reference
84	Po	145	7
85	At	—	
86	Rn	—	
87	Fr	—	
88	Ra	—	
89	Ac	—	
90	Th	597	8
91	Pa	569	8
92	U	531	8
93	Np	465	8
94	Pu	345	8
95	Am	284	8
96	Cm	387	8
97	Bk	310	8
98	Cf	196	8
99	Es	133	8
100	Fm	143	9

References

1. I. F. Silvera, Rev. Mod. Phys. **52**, 393 (1980).
2. R. K. Crawford, in *Rare Gas Solids*, M. L. Klein and J. A. Venables, eds. (Academic, London, 1977) vol. 2, chap. 11.
3. L. V. Gurvich, V. S. Yorish, N. E. Khandamirova, and V. S. Yungman, in *Handbook of Thermodynamic and Transport Properties of Alkali Metals*, R. W. Ohse, ed. (Blackwell, Oxford, 1985) chap. 6.6.2.
4. R. Hultgren, P. D. Desai, D. T. Hawkins, M. Gleiser, K. K. Kelley, and D. D. Wagman, *Selected Values of the Thermodynamic Properties of the Elements* (Am. Soc. Metals, Metals Park, Ohio, 1973).
5. T. A. Scott, Phys. Repts. **27**, 89 (1976)
6. G. E. Jelinek, L. J. Slutsky, and A. M. Karo, J. Phys. Chem. Solids **33**, 1279 (1972).
7. K. A. Gschneidner, Jr., Solid State Phys. **16**, 275 (1964).
8. J. W. Ward, P. D. Kleinschmidt, and D. E. Peterson, in *Handbook on the Physics and Chemistry of the Actinides*, A. J. Freeman and C. Keller, eds. (North-Holland, Amsterdam, 1986) vol. 4, chap. 7.
9. R. G. Haire and J. K. Gibson, J. Chem. Phys. **91**, 7085 (1989).

APPENDIX E
Melting Temperature

The listed temperatures are the 0.1 MPa melting temperatures or, where the triple-point pressure exceeds 0.1 MPa, as in C and As, the triple-point temperatures.

Z	Element	Melting Temperature (K)	Reference
1	H	13.80	1
1	D	18.69	1
2	He-3	(liquid)	
2	He-4	(liquid)	
3	Li	453.7	2
4	Be	1562	2
5	B	2365	2
6	C	4530	3
7	N	63.15	2
8	O	54.36	2
9	F	53.48	2
10	Ne	24.56	2
11	Na	371	2
12	Mg	923	2
13	Al	934	2
14	Si	1687	2
15	P	900	4
16	S	388	2
17	Cl	172.2	2
18	Ar	83.80	2

(continued)

Appendix E (continued)

Z	Element	Melting Temperature (K)	Reference
19	K	337	2
20	Ca	1115	2
21	Sc	1814	2
22	Ti	1943	2
23	V	2183	2
24	Cr	2136	2
25	Mn	1519	2
26	Fe	1811	2
27	Co	1768	2
28	Ni	1728	2
29	Cu	1358	2
30	Zn	693	2
31	Ga	303	2
32	Ge	1212	2
33	As	1090	5
34	Se	494	2
35	Br	266	2
36	Kr	115.8	2
37	Rb	313	2
38	Sr	1042	2
39	Y	1795	2
40	Zr	2128	2
41	Nb	2742	2
42	Mo	2896	2
43	Tc	2428	2
44	Ru	2607	2
45	Rh	2236	2
46	Pd	1828	2
47	Ag	1235	2
48	Cd	594	2
49	In	430	2
50	Sn	505	2
51	Sb	904	2

(continued)

Appendix E (continued)

Z	Element	Melting Temperature (K)	Reference
52	Te	723	2
53	I	387	2
54	Xe	161.4	2
55	Cs	302	2
56	Ba	1002	2
57	La	1191	2
58	Ce	1071	2
59	Pr	1204	2
60	Nd	1294	2
61	Pm	1315	2
62	Sm	1347	2
63	Eu	1095	2
64	Gd	1586	2
65	Tb	1629	2
66	Dy	1685	2
67	Ho	1747	2
68	Er	1802	2
69	Tm	1818	2
70	Yb	1092	2
71	Lu	1936	2
72	Hf	2504	2
73	Ta	3293	2
74	W	3695	2
75	Re	3459	2
76	Os	3306	2
77	Ir	2720	2
78	Pt	2042	2
79	Au	1338	2
80	Hg	234	2
81	Tl	577	2
82	Pb	601	2
83	Bi	545	2

(continued)

Appendix E (continued)

Z	Element	Melting Temperature (K)	Reference
84	Po	527	2
85	At	—	
86	Rn	202	2
87	Fr	—	
88	Ra	973	2
89	Ac	1324	2
90	Th	2028	2
91	Pa	1845	2
92	U	1408	2
93	Np	912	2
94	Pu	913	2
95	Am	1449	2
96	Cm	1618	2
97	Bk	1323	2
98	Cf	1173	2
99	Es	1133	2
100	Fm	—	

References

1. P. C. Souers, *Hydrogen Properties for Fusion Energy* (University of California Press, Berkeley, Los Angeles, London, 1986) p. 126.
2. Anonymous, Bull. Alloy Phase Diag. **7**, 601 (1986).
3. A. Cezairliyan and A. P. Miiller, Bull. Am. Phys. Soc. **32**, 608 (1987).
4. J. F. Cannon, J. Phys. Chem. Ref. Data **3**, 781 (1974).
5. K. A. Gschneidner, Jr., Solid State Phys. **16**, 275 (1964).

PERIODIC TABLE

1 H																	2 He
3 Li	4 Be											5 B	6 C	7 N	8 O	9 F	10 Ne
11 Na	12 Mg											13 Al	14 Si	15 P	16 S	17 Cl	18 Ar
19 K	20 Ca	21 Sc	22 Ti	23 V	24 Cr	25 Mn	26 Fe	27 Co	28 Ni	29 Cu	30 Zn	31 Ga	32 Ge	33 As	34 Se	35 Br	36 Kr
37 Rb	38 Sr	39 Y	40 Zr	41 Nb	42 Mo	43 Tc	44 Ru	45 Rh	46 Pd	47 Ag	48 Cd	49 In	50 Sn	51 Sb	52 Te	53 I	54 Xe
55 Cs	56 Ba	57 La	72 Hf	73 Ta	74 W	75 Re	76 Os	77 Ir	78 Pt	79 Au	80 Hg	81 Tl	82 Pb	83 Bi	84 Po	85 At	86 Rn
87 Fr	88 Ra	89 Ac															

58 Ce	59 Pr	60 Nd	61 Pm	62 Sm	63 Eu	64 Gd	65 Tb	66 Dy	67 Ho	68 Er	69 Tm	70 Yb	71 Lu
90 Th	91 Pa	92 U	93 Np	94 Pu	95 Am	96 Cm	97 Bk	98 Cf	99 Es	100 Fm	101 Md	102 No	103 Lr

ABBREVIATIONS

Abbreviations Used in This Book

AIP	*Ab Initio* Pseudopotential
APW	Augmented Plane Wave
ASA	Atomic-Sphere Approximation
ASW	Augmented Spherical Wave
ca.	circa
CDW	Charge-density wave
cp	close-packed
DAC	Diamond-Anvil Cell
DTA	Differential Thermal Analysis
EOS	Equation of State
FPLAPW	Full-Potential Linear Augmented Plane Wave
GPT	Generalized Pseudopotential Theory
KKR	Korringa-Kohn-Rostocker
LAPW	Linear Augmented Plane Wave
LCGTO	Linear Combination of Gaussian-Type Orbitals
LDA	Local-Density Approximation
LMTO	Linear Muffin-Tin Orbitals
LSDA	Local Spin-Density Approximation
LV	Liquid-Vapor
MC	Monte Carlo
MD	Molecular Dynamics
MT	Muffin Tin
NFE	Nearly Free Electron
NMR	Nuclear Magnetic Resonance
OCP	One-Component Plasma
QMC	Quantum Monte Carlo
RP	Room Pressure (1 atm)
RT	Room Temperature (300 K)
RTP	Room Temperature and Pressure (300 K and 1 atm)
XRD	X-Ray Diffraction
vs.	versus

INDEX

Designer:	UC Press Staff
Text:	10/12 Baskerville
Display:	Baskerville
Printer:	Bookcrafters, Inc.
Binder:	Bookcrafters, Inc.